Environmental Movements and Waste Infrastructure

As rates of consumption grow, so has the problem of waste management. National and local waste authorities seek to manage such problems through the implementation of state regulation and construction of waste infrastructure, including landfills and incinerators. These, however, are undertaken in a context of increasing supra-state regulatory frameworks and directives, and of increasing activity by multi-national corporations, and are increasingly contested by activists in the affected communities.

Environmental Movements and Waste Infrastructure examines the structures of political opportunity that confront environmental movements that challenge the state or corporate sector. Case studies on collective action campaigns from the EU, US and Asia illuminate the similarities and differences between protests against waste incinerators and other waste management infrastructure projects within different states. Several contributions share a concern about cross-border or trans-national waste flows. Each case study looks beyond its initial local frame of reference and interrogates assumptions about NIMBYism or localism, demonstrating the wider linkages and networks established by both grassroots campaigns and state and multinational agencies

This book was previously published as a special issue of *Environmental Politics*.

Christopher Rootes is Professor of Environmental Politics and Political Sociology as well as Director of the Centre for the Study of Social and Political Movements at the University of Kent, Canterbury.

Liam Leonard is Lecturer in Sociology, Criminology and Human Rights at the Institute of Technology, Sligo, Ireland.

Environmental Movements and Waste Infrastructure

Edited by Christopher Rootes and Liam Leonard

Routledge
Taylor & Francis Group

LONDON AND NEW YORK

First published 2010 by Routledge
2 Park Square, Milton Park, Abingdon, Oxon, OX14 4RN

Simultaneously published in the USA and Canada
by Routledge
711 Third Avenue, New York, NY 10017

Routledge is an imprint of the Taylor & Francis Group, an informa business

First issued in paperback 2012

This book is a reproduction of *Environmental Politics*, vol. 18, issue 6. The Publisher
requests to those authors who may be citing this book to state, also, the bibliographical
details of the special issue on which the book was based

Typeset in Times by Value Chain, India

British Library Cataloguing in Publication Data
A catalogue record for this book is available from the British Library

ISBN13: 978-0-415-45869-6 (HBK)
ISBN13: 978-0-415-81476-8 (PBK)

CONTENTS

Notes on contributors

Iosif Botetzagias is a Lecturer in Environmental Politics and Policy, Department of Environment, University of the Aegean, Greece. Publications include 'The influence of social capital on willingness to pay for the environment', *European Societies*, 11 (4) (2009) and 'Accounting for difficulties faced in materializing a Transnational ENGO Conservation Network' *Global Environmental Politics*, 10 (1) (2010).

Peter Doran is a Lecturer in sustainable development, environment and planning in the School of Law, Queen's University Belfast, Northern Ireland. A senior editor and writer on the *Earth Negotiations Bulletin* for the International Institute for Sustainable Development, his publications include: 'The Global Justice Movement and Sustainable Development', *International Journal of Green Economics* 1(1) 2006.

Honor Fagan is a Lecturer in Sociology at the National University of Ireland Maynooth. Her publications include: Globalization and Human Security: An Encyclopaedia (Praeger, 2008) (co-edited with R. Munck) and (co-authored with M. Murray) 'Green Ireland? Waste in its Social Context' in B. Bartley and R. Kitchin (eds.) *Understanding Contemporary Ireland* (Pluto, 2007).

John Karamichas is a lecturer in Sociology in the School of Sociology, Social Policy and Social Work at Queen's University Belfast, Northern Ireland. Recent publications include 'Accessing the Institutions: The Road to the Socialist-Green Alliance in Spain', *Mediterranean Politics* 13(3), (2008).

Melissa Kemberling (Toffolon-Weiss) is a Senior Epidemiologist at the Alaska Native Tribal Health Consortium in Anchorage, Alaska. Her publications include *Chronicles from the Environmental Justice Frontline* (Cambridge University Press, 2001) (co-author), and 'Alaska Native Parental Attitudes on Cervical Cancer, HPV and the HPV Vaccine', *International Journal for Circumpolar Health* 67(4) (2008).

Su-ming Khoo is a Lecturer in the School of Political Science and Sociology, National University of Ireland, Galway. Her publications include 'Globalization, terror and the future of "development"' in M. Mullard (ed.) *Globalization, Citizenship and the War on Terror* (Edward Elgar, 2007), and 'Development Education, Citizenship and Civic Engagement', *Policy and Practice*, 1 (3) (2006).

Liam Leonard is a Lecturer in Sociology at the Institute of Technology, Sligo, Ireland. Publications include *Politics Inflamed: Campaigns against Incineration in Ireland* (Greenhouse Press, 2005), *Green Nation* (Greenhouse Press, 2006), *The Environmental Movement in Ireland* (Springer, 2008) and *The Transition to Sustainable Living and Practice* (ed., with John Barry) (Emerald).

Darren McCauley is a Lecturer in European and Environmental Politics at Queen's University Belfast, and a research associate of the Queen's Institute for a Sustainable World. His contribution was partly written during his time as Lecturer in Environmental Geography at Trinity College Dublin.

Henrike Rau is a lecturer in Political Science and Sociology at the National University of Ireland, Galway. Her publications include: *Environmental Argument and Cultural Difference* (ed., with R. Edmondson) (Peter Lang, 2008).

Timmons Roberts is Professor of Sociology and Environmental Studies and Director of the Center for Environmental Studies, Brown University, USA. Co-authored publications include *Chronicles from the Environmental Justice Frontline* (Cambridge University Press, 2001), *A Climate of Injustice: Global Inequality, North-South Politics, and Climate Policy* (MIT Press, 2007), and *Greening Aid?* (Oxford University Press, 2008).

Christopher Rootes is Professor of Environmental Politics and Political Sociology and Director of the Center for the Study of Social and Political Movements, University of Kent, Canterbury, England. His publications include *Environmental Movements: local, national and global* (ed.) (Cass, 1999); *Environmental Protest in Western Europe* (ed.) (Oxford University Press, 2003/7) and *Acting Locally: Local environmental mobilizations and campaigns* (ed.) (Routledge, 2008).

Environmental movements, waste and waste infrastructure: an introduction

Christopher Rootes

Centre for the Study of Social and Political Movements, School of Social Policy, Sociology and Social Research, University of Kent at Canterbury, UK

The increasing amount and complex nature of municipal waste presents problems of management. Recognising the inadequacies of landfill, waste management authorities proposed incineration, but large-scale incineration provoked more public concern and protest. Concerns about toxicity of incinerator emissions led to tighter regulation, but as evidence of the impacts of air pollution upon human health has hardened, opposition to incineration has persisted. The inequitable distribution of exposure to waste-related risks has generalised demands for environmental justice. There is variation in the extent to which anti-incinerator campaigns are networked among themselves and with environmental NGOs, but such networking has increased and is now transnational. New technologies mitigate some of the hazards of modern waste management but are unlikely to eliminate public protest over the siting of waste infrastructure.

Waste is, as the then Director of the United States Environmental Protection Agency, William Ruckelshaus declared in 1972, 'a fundamental ecological problem', illustrating, perhaps more clearly than any other, the need for change in traditional attitudes and practices (quoted by Luton 1996, p. 259). It is, however, a problem that crept up on industrialising and industrialised societies, and it is only recently that it has stimulated widespread popular mobilisations. The urban sanitation 'movements' that arose in the nineteenth century in various countries were almost invariably elite initiatives rather than grassroots movements. The conservationist and preservationist movements that developed from the late nineteenth century onwards were preoccupied with the impact upon the natural environment of industrialisation and urbanisation and, in the New World, of land clearance for agricultural and pastoral activities. It was only with the emergence, in the 1960s and 1970s, of reform environmentalism that environmental movements emerged that focused upon problems of the urban environment and of the impacts of industrialism upon human health and well-being. Only then did waste become a major item on the environmental movement's agenda.

Environmental movements

It is perhaps necessary to clarify at the outset what I mean by an 'environmental movement'. Adapting what Diani (1992) describes as the then emerging consensual definition of a social movement, an environmental movement may be defined as a loose, non-institutionalised network of organisations of varying degrees of formality, as well as individuals and groups with no organisational affiliation, that are engaged in collective action motivated by shared identity or concern about environmental issues (Rootes 2004, p. 610). The existence of a network is as crucial to the definition as is collective action; it is perfectly possible to have many individual, isolated but vigorous protests that attract large numbers of people, but so long as they are unconnected one to another, they do not constitute a social movement.

The advantage of the network approach is that whether protests, protesters or organisations constitute a movement is an empirical question to be settled by scrutiny of the network links, collective action and evidence of shared identity. Networking, collective action and shared concern are best conceived of as matters of degree or points on a continuum: the forms and intensity of both action and concern, and the degree of integration of the network, may vary considerably from place to place and from time to time.

We shall, in this volume, encounter environmental movements of various kinds in the several countries with which our contributors deal. It will quickly be apparent that whilst there have been campaigns of some intensity against waste infrastructure, there is significant variation from one country to another; not all of these campaigns have been networked into movements, and their relationship with broader environmental movements has differed considerably. But before we turn our attention to the cases considered later in this volume, it may be helpful to consider the nature of waste, the various means by which it has been managed, and the reasons that it has become the subject of so much contention.

Waste and waste management

Article 1(a) of the EU Waste Framework Directive defines waste as 'any substance or object ... which the holder discards or intends or is required to discard'. It helpfully goes on: 'Whether or not a substance is discarded as waste – and when waste ceases to be waste – are matters that must be determined on the facts of the case' (http://www.defra.gov.uk/environment/waste/topics/index.htm, accessed 12 October 2008). The European Commission's difficulty in determining what is or might be considered waste is illustrative of the uncertainties that, despite the welter of seemingly solid statistics and time series, surround all discussions of waste. Waste is messy and changeable stuff.

Waste comes in many forms. Some of it, especially that labelled 'hazardous', is self-evidently controversial and is correspondingly likely to excite public concern almost anywhere. More surprising, perhaps, is the way in which waste that is not so labelled and whose disposal was for generations generally regarded as unproblematic has in recent decades become increasingly controversial.

The tale most often told about the 'waste crisis' is that it is a product of our increasingly wasteful consumer society. Compared to our forebears and to people in less affluent societies, we buy more things that are more lavishly packaged and more quickly discarded, with the result that the amount of waste so generated becomes a problem of unprecedented proportions (see, e.g., Tammemagi 1999, p. 25). But as O'Brien (2007) argues, the amount of waste we produce is less problematic than the changing composition of that waste. In fact, such are the changes in the nature of waste produced by Western households over the past century that it is difficult even to know whether households are, in aggregate, producing more waste. The alleged increases in the amounts of waste requiring management and disposal may simply reflect changes in what households are expected to dispose of themselves. It is likely, for example, that as households were discouraged from burning waste on open fires or in backyards, so the amounts of waste entering the recorded waste stream increased. Similarly, as kerbside collections of garden waste or yard trimmings have been provided, so the weight of waste collected has risen, as has the weight of waste 'recycled or composted'.

Yet if it is not clear that the *amount* of post-consumer waste has greatly increased, it *is* clear that, with technological innovation there has been a dramatic increase since World War II in the variety, availability and use of synthetic chemicals and complex compound materials. The nature of the products that people consume has changed and so, as a consequence, has the nature of the waste they discard. As O'Brien (2007, p. 85) puts it: 'there has been a proliferation in the variety of hazardous ingredients that make up contemporary consumer products and in the routes by which these ingredients reach the environment both during and after consumption. Post-war industrial development has indeed been characterised by increases in the toxic mix and hazard of the ingredients of consumer goods and post-consumption wastes'. Furthermore, advances in scientific knowledge have made scientists, citizens and policymakers alike aware of the possible hazards of waste that was in the past regarded as benign. The upshot is that the safe and publicly acceptable management of post-consumer waste has become much more complicated.

Most of the contributions to this volume focus upon the means of treatment of that form of waste which is most familiar, apparently relatively innocuous, but in practice especially troublesome: municipal solid waste (MSW), most of which is generated in and by households. MSW comprises only a relatively small proportion of the total quantities of waste generated in industrialised societies – less than 25% of all industrial, commercial and municipal waste in England, for example. Nevertheless, disposing of it presents special difficulties for several reasons.

First, as societies become more affluent and, especially, as urban populations have increased, so the total amount of waste has tended to rise and the scale of the waste management problem has increased accordingly. Moreover, as the contents of household waste have changed as industrial processes and lifestyles have changed, so forms of waste disposal once considered appropriate or adequate have lately become problematic.

Compared with industrial waste, MSW is a complex mix of locally and seasonally variable materials. Ideally, MSW might be separated and treated in the manner most appropriate to the chemical constituents of its various components; thus biodegradable food scraps and garden waste would be dealt with differently from paper and cardboard, which in turn would be separated from plastics, metals, wood and glass. Even where householders can be persuaded to separate their waste at source, they may not do so reliably, and so a complex problem is passed on to waste disposal.

Industrial waste is not, of course, unproblematic, but by comparison with households, industry recycles a higher proportion of the waste it produces, waste that is in any case more homogeneous and of known character in quantities large enough to justify sorting. Moreover, because waste is easily recognised as a cost, so industry has clear economic incentives to minimise waste and/or to recycle. Thus, in England and Wales in 1998–99, 44% of industrial waste was recycled or composted, compared with 24% of commercial waste and just 12% of household waste (Strategy Unit 2002). Even the tripling by 2008 of the proportion of municipal waste recycled does not bring it up to the rate of industrial waste recycling a decade earlier.

Since no-one wants to live next door to a waste facility, their siting has become increasingly problematic, especially as population growth and urbanisation have increased population densities. As a consequence, the places where MSW is generated and from which it must be collected are widely dispersed. If waste is to be disposed of safely and appropriately in places where it will not cause nuisance to neighbours, it has to be transported, and if it is treated in sophisticated plants that require large volumes to achieve economies of scale, the distances over which it is transported may be considerable, and the vehicle traffic so generated may itself create environmental as well as management problems.

Managing waste

All forms of waste disposal may be controversial. Indeed, the most nearly universal means of disposal of MSW – burying it in holes in the ground – has become especially problematic. This is for three reasons. First, the immediate environmental impacts of ill-managed rubbish dumps and landfills, with their attendant stenches and vermin ranging from rats and flies to seagulls, was increasingly apparent the larger and more ubiquitous those landfills became. Second, the toxicity of the leachate produced by such landfills as they became saturated by rainfall or reached the water table or streams grew more apparent, initially because of the local fish kills and deaths of crops and trees, and concerns about it were exacerbated as the persistent character of the toxins so distributed were better understood. Third, and perhaps most compelling for policymakers, was the accelerating progress toward full capacity of existing landfill sites, many of which were voids left as by-products of previous extraction of minerals and building materials.

The impending exhaustion of existing landfill capacity led to an increasingly urgent search for new landfill sites just as increasing scientific knowledge of the hazards of landfill – 'uncontrolled biochemical reactors' as Krohn and van den Daele (1998) called them – and better understanding of the minimum geological requirements for safe landfill limited the possible options.[1] The search for new landfill sites was made more difficult by increasing public opposition, both from those who had experience of the downsides of existing or older poorly managed sites (extensions of existing sites being the option most widely chosen) and from those, including countryside conservation organisations as well as local amenity societies and campaign groups, who sought to defend existing uses of the environment and its amenity value from such unwelcome development.

But it is especially incineration, rather than other forms of management or disposal of municipal waste, that has most often excited oppositional mobilisations. The principle reason for this is that fear of emissions to air is more nearly universal than fear of the other forms of pollution that may attend waste management. Compared with pollution of water or soils, air pollution is more capricious because its distribution depends upon the strength and direction of prevailing winds, and it is more difficult to avoid, especially because it is so often invisible.

Waste incineration

High temperature incineration has been the generally recommended means of disposing of many forms of toxic waste. It is also the surest way of dispersing toxins into the air that we breathe. Emissions from the best-designed and best-managed incinerators can, in principle, be monitored, managed and contained to levels deemed safe by regulatory authorities and by prevailing scientific consensus. But, as scientific knowledge and the technological capacity for measuring pollutants and their effects have grown, so what are considered safe levels of emissions have shown an alarming tendency to change over time, invariably downwards.

Moreover, with ever more sophisticated manufacturing processes involving increasingly complex compounds, materials unknown a generation ago enter the waste stream, the complexity of waste increases, and the need arises to measure and minimise the emission of more and more previously unmeasured and unmonitored substances, including products of the combustion process that are potentially more toxic than the original waste.

Unfortunately, even the best-designed and best-managed incinerators sometimes go wrong. Worse, some are neither well designed nor reliably well managed. In the bland parlance of the trade, 'emissions exceedances' are quite routine, and accidents, occasionally catastrophic, occur, sometimes in the full glare of publicity, too often in the twilight of non-transparency that in many places surrounds strategically important but actually or potentially hazardous industrial processes.

It is scarcely surprising, then, that waste incinerators have become unwelcome neighbours, and that the shadow of suspicion falls especially heavily over incinerators designed to dispose of toxic waste, particularly where that waste is trucked in from other places. In the course of the last three decades, as awareness of the attendant risks has spread, so communities have become increasingly resistant to the siting of new waste incinerators in their vicinity. Some of these local campaigns have become epic struggles and their stories have become scarcely credible tales of corporate and official deception and skulduggery that are remarkably revealing about the distribution and exercise of power in the modern world (see, e.g., Shevory 2007).

Municipal waste may not be flagged as hazardous; it is, at least as collected, popularly regarded as relatively innocuous, but many of the same concerns exist about emissions from the incineration of MSW as from that of waste labelled 'toxic'.

Health risks

There are essentially two methods by which the actual or potential impacts of incinerators upon human health may be calculated: by measuring and assessing the toxicity of the emissions from incinerator stacks; or by epidemiological studies of people living or working in the vicinity of incinerators. Actual harm is extremely difficult to demonstrate, partly because it may take many years for a particular pathology to become apparent and to be diagnosed, and also because so many confounding variables intervene to prevent epidemiological studies from approximating to the standards of experimental laboratory research in the identification of a specific source of the pathology in question. Even in the cases of the pathologies considered most likely to be triggered by incinerator emissions – respiratory illness and certain cancers – the evidence is sparse. However, although it is difficult or impossible to demonstrate the harmful effects of incinerator emissions, the absence of clear evidence of actual harm does not warrant the conclusion that there can be certainty about the absence of actual or potential harm.

The 1999 report of the US National Research Council (2000) concluded that, when operated properly by well-trained operatives, modern waste incinerators posed little risk to public health. But it conceded that older designs, human error, and equipment failure could result in higher than normal short-term emissions that needed further study. Few investigations had tried to establish a link between an incinerator and illness in the surrounding area, and most had been unable to detect any adverse health effects. A decade of further research has done little to qualify the 1999 assessment (see, e.g., Saffron *et al.* 2003, HPA 2009).

While older waste incineration plants emitted unacceptably high levels of pollutants, recent regulatory changes and new technologies have significantly reduced them. In the US, EPA regulations in 1995 and 2000 under the Clean Air Act succeeded in reducing emissions of dioxins from waste-to-energy facilities by more than 99% below 1990 levels, while mercury emissions reduced by over 90% (Psomopoulos *et al.* 2009). The EPA noted these improvements in 2003, citing waste-to-energy as having 'less environmental impact than almost any other source of electricity' (http://www.wte.org/userfiles/file/epaletter.pdf, accessed 4 September 2009).

The evidence of demonstrable harm caused by modern incinerators may be elusive, but perception is all, and the public fear of incinerator emissions has not abated even though the technology has improved and emissions are cleaner. Initially, fears focused especially upon dioxins, of which waste incinerators were, in the US, the Netherlands and Britain, the major source up to the mid-1990s. Dioxins are a product of most forms of combustion, but especially of the combustion of chlorinated materials, including chlorine-bleached paper, even trace amounts of which appear to be sufficient to produce dioxins. Since the mid-1990s dioxins have been very effectively controlled in licensed incinerators, and levels of dioxins do not now generally appear to be higher in the vicinity of waste incinerators than elsewhere. Indeed, the Chlorine Chemistry Division of the American Chemistry Council (2003) estimated that by 2004, the majority of all dioxin emissions in the US would be attributable to backyard trash burning, emissions from which would far exceed those from the aggregated emissions of all waste incinerators.

As concerns about dioxin emissions have been countered by increasingly tight regulation and lower levels of emission, so concern has shifted to the particulate content of emissions, and especially of fine particulate matter ($PM_{2.5}$), ultrafine particles (PM_1) and nanoparticles ($PM_{0.1}$). These microscopic particles, which are emitted by many forms of combustion, pose health risks due to their ability to pass unfiltered through the nose and mouth, penetrating deep into human lungs and bloodstreams, where they can cause potentially fatal respiratory and/or pulmonary diseases.

Emissions of both dioxins and fine particles from waste incinerators are now exceeded by those from road traffic, even though vehicle emissions are themselves much cleaner now than they were two decades ago. In the EU, $PM_{2.5}$ emissions are still not routinely or systematically measured in all states (and PM_1 and $PM_{0.1}$ emissions are scarcely monitored at all), but the EU's survey of air quality in 2007 identified residential sources as contributing 27% of $PM_{2.5}$ pollution, considerably ahead of private cars (6%) and heavy transport vehicles (5%) (EEA 2009). Emissions from incinerators were not separately identified. In the UK, incinerator emissions are estimated to make only a very small contribution (0.03%) to aggregate emissions of particulates by comparison with traffic (27%) and industry (25%), and, based on one modelling exercise by the Environment Agency, the additional contribution they make to local concentrations of particles appears to be very small (HPA 2009).

Scientific understanding of the impacts of air pollution continues to advance. The UK Committee on the Medical Effects of Air Pollutants (COMEAP), basing its review on new research showing hardening evidence of harm, in 2009 concluded that 'the evidence as a whole points strongly to an association between long-term exposure to particulate air pollution and effects on mortality' (COMEAP 2009, p. 1). Accordingly, COMEAP dramatically increased its estimates of the mortality rates attributable to particulate pollution. In 2001, in its first report, COMEAP suggested policymakers should assume that a 10-microgram-per-cubic-metre increase in fine particulate matter ($PM_{2.5}$) increases overall mortality rates by 1–9%, with most committee members estimating actual increases closer to 1%. In 2009, however, COMEAP advised that policymakers should assume increased mortality of 1–12%, with a best estimate of 6%.

Alarming as it is, COMEAP's report has been criticised for not taking account of studies published after early 2006 that suggest even higher risk rates (*ENDS Report* 414, July 2009, p. 21). The six-city study undertaken for the American Cancer Society revealed a linear relationship between $PM_{2.5}$ concentrations and mortality (Laden *et al.* 2006). A 36-city study of women's health found strong correlations between mortality and long-term exposure to $PM_{2.5}$ concentrations (Miller *et al.* 2007). More wide-ranging reviews of the literature also suggest significant impacts of $PM_{2.5}$ pollution upon mortality and chronic disease, especially pulmonary and cardiac pathologies (Pope and Dockery 2006; Chen *et al.* 2008).

The European Commission estimates that 350,000 of the 370,000 premature deaths annually attributable to air pollution are attributable to fine particles, and the EU Air Quality Directive (2008) for the first time set EU-wide limits on $PM_{2.5}$ emissions. However, the limits prescribed by the Directive apply only from 2010 'if possible' or 2015 (http://www.euractiv.com/en/environment/eu-adopts-stricter-air-quality-rules/article-171631). Since fine particle emissions are still not routinely measured in much of Europe, it is unlikely that progress will be swift, but these new standards will inform decisions concerning new waste facilities.

Further research is required in order to identify which components of $PM_{2.5}$ are most damaging, but it is apparent that fine particles have adverse health effects at levels well below those permitted by the EU Directive (Cassee 2009). Thus the concerns of anti-incinerator campaigners about the health impacts of emissions to air are in this respect increasingly vindicated by scientific evidence.

There are good reasons to doubt that the impacts of incinerators upon human health and environmental quality are adequately understood let alone reflected in existing emissions standards; the progressive tightening of those standards is a response to mounting scientific knowledge of actual and potential harm rather than an increasingly nervous invocation of the precautionary principle (for a succinct and cogent discussion of the issues, see Shevory 2007, pp. 154–169).

Yet because the detriments to air quality attributable to the most commonly measured pollutants emitted by modern waste incinerators are much lower than those associated with earlier generations of incinerators, the UK Health Protection Authority (2009), in its most recent review of research examining the suggested links between emissions from municipal waste incinerators and effects on health, concluded:

> While it is not possible to rule out adverse health effects from modern, well regulated municipal waste incinerators with complete certainty, any potential damage to the health of those living close-by is likely to be very small, if detectable. This view is based on detailed assessments of the effects of air pollutants on health and on the fact that modern and well managed municipal waste incinerators make only a very small contribution to local concentrations of air pollutants. The Committee on Carcinogenicity of Chemicals in Food, Consumer Products and the Environment has reviewed recent data and has concluded that ... any potential risk of cancer due to residency near to municipal waste incinerators is exceedingly low and probably not measurable by the most modern techniques. (HPA 2009, p. 1)

Like the US National Research Council a decade earlier, the HPA (2009, p. 2) suggests that such health impacts as can be demonstrated are attributable to 'older incinerators with less stringent emission standards and cannot be directly extrapolated with any reliability to modern incinerators'.

Not the least problem with these repeated reassurances is that, although the advice that incineration is safe is essentially the same, the levels of emissions that are considered safe have been steadily reduced over time as new scientific evidence of the toxicity of the components of emissions has accumulated, and as epidemiological studies have tended to produce firmer evidence of the adverse impacts of air pollution upon human health.

But even if waste incinerators make a very small contribution to air pollution, and even if, as the HPA concludes, modern, well managed incinerators make only a small contribution to local concentrations of air pollutants, they are, actually or potentially, a point source of pollution that is perceived as threatening by their prospective neighbours. 'Worrying about such things is in itself one of the external costs imposed upon communities' (Shevory 2007, p. 210). Given the way in which thresholds below which emissions are considered safe have been progressively lowered in the light of better measurement and new evidence, such public anxieties cannot be dismissed as unreasonable.

Since most people who generate waste do not live near the sites chosen for incinerators, questions of equity are raised. Waste incinerators are often proposed for areas in which there is already a concentration of social deprivation and of environmental pollutants that may be presumed to have negative effects upon human health. Some studies do suggest that there are higher rates of cancer or cardiac disease among the neighbours of municipal waste incinerators, but the general conclusion of the various ecological studies of the impact of municipal waste incinerators is that it is difficult to demonstrate any impacts upon the health of the people living nearby, very probably because the specific effects of incinerator emissions are confounded by different factors, environmental or other. For that reason, even observable increases in pathology are difficult or impossible to distinguish from the effects of other factors in urban environments. But the implication of this is surely not that the siting of new waste incinerators in such areas is of no consequence; on the contrary, it suggests that in areas where the health of the population is already impaired, for whatever reason, it is at best inequitable to add to their burdens, even if the best guess is that the additional burdens and the health impacts specifically attributable to them are very small or not presently detectable.

Such arguments have been marshalled in the course of resistance to waste incinerator sitings in the US and in Britain, as well as by opponents even of highly engineered sanitary landfills, as in Greece (see Botetzagias and Karamichas, this volume). They reflect the general contention of the environmental justice movement that the concentration of environmental 'bads' in a small area imposes unreasonable burdens upon those who live there. Indeed, the distributive inequities of environmental risks may lead to claims that 'enough is enough' even where no claims about adverse health impacts are made (Schlüter *et al.* 2004, p. 728).

Contention over waste

Political contention over matters of waste is widespread in industrialised societies, and in some less industrialised ones. But whilst grassroots campaigns, with or without support from more formally organised environmental movement organisations and NGOs, are common, they vary considerably in their incidence and intensity, the issues they take up, the forms of their discourses, the extent to which they cohere as movements, and their outcomes. We shall in this volume focus attention upon a small slice of that rich diversity.

The first case considered here is that of the USA. Campaigns against waste infrastructure in the US emerged in the 1970s against a background of increasing public anxiety about the impacts of industrialism and industrialised agriculture upon the environment and human health, and at a time when protests against the proposed sitings of nuclear power stations were approaching their peak. Both began in widespread grassroots protests, and waste campaigners, focused upon landfills and hazardous waste, quickly raised issues of the inequitable distribution of waste dumps. Building upon the legacies of the civil rights movement, the emergent environmental justice movement was the most significant innovation in US environmentalism since the advent of reform environmentalism at the end of the 1960s. It helped to sustain and to ensure the networking of anti-incineration campaigns during the 1980s and, in due course, together they contributed to the reinvigoration of the US environmental movement (Rootes and Leonard, this volume).

There are conspicuous differences between the US and English cases. In the US, the established organisations that are generally held to comprise the US environmental movement played at best marginal roles in campaigns against waste infrastructure, and there it was new organisations that emerged from local grassroots campaigns themselves that achieved the networking of local campaigns that enables us to speak confidently of anti-waste movements in the US. In England, by contrast, local campaigners, left to themselves, struggled to achieve the degree of networking that would justify their being called a movement, and here it was the interventions of national environmental organisations, notably Friends of the Earth (FoE), that fostered the belated development of an anti-incineration network (Rootes, this volume).

The US and English cases are, however, similar insofar as, in both cases, the declining appeal of waste incineration was not simply a product of anti-incineration campaigns. In both countries, the fluctuating fortunes of incineration have been strongly influenced by government policies and, especially in the US, by the shifting economics of waste-to-energy (or, as it is perhaps more beguilingly termed in England, energy-from-waste). Where the US and England differ, however, is in policies toward landfill. Perhaps surprisingly in view of the way in which protests over the landfilling of hazardous wastes stimulated the mobilisation of the environmental justice movement, it is the expansion of sanitary landfills that has undermined the economic case for waste incineration in the US. In England, by contrast, it was the imperative to reduce the proportion of waste sent to landfill that promised a new dawn for waste incineration.

Four of the contributions to this volume deal with movements or campaigns against waste infrastructure projects in EU member states. All are subject to the same developing regime of EU environmental Directives, including in particular those concerning waste and the means of its disposal. But these four states are differently positioned with respect to waste issues, have developed different patterns of policy response, and have very varied experience of citizen campaigns concerning waste. France embarked upon an energetic policy of diversion of waste from landfill in the 1970s and has been a leader in the development of a network of waste incinerators and a strong advocate of waste incineration within the EU (see McCauley, this volume). Britain, Greece and Ireland, by contrast, are all countries that at the start of the present century relied upon landfill as the principal means of disposing of their municipal waste and so, in order to comply with the EU Landfill Directive, they have been obliged, with increasing urgency, to develop alternatives. But whereas British and Irish waste authorities have been driven to consider incineration, Greece, having ruled out incineration as a risk to health, has chosen instead to attempt to develop a network of waste treatment facilities that do not involve incineration but are essentially highly engineered landfills (see Botetzagias and Karamichas, this volume).

Yet, interestingly, in Greece waste has given rise to a higher proportion of nationally reported environmental protests than in any of the European countries for which we have systematic evidence (Kousis 2003, pp. 117–119, 2007). In Ireland, protests involving waste disposal constituted the largest single thematic cluster of all environmental protests during the 1990s (Garavan 2004, p. 79). In Britain, by contrast, in only one year during the decade 1988– 1997 did nationally reported protests concerning waste rise above low single figures (Rootes 2003, p. 30). Thus Britain, despite its heavy reliance on landfill, had a low incidence of reported protests concerning waste, a pattern more similar to that of other Northern industrialised countries than to that of countries such as Greece, Italy, Spain or Ireland that also relied upon landfill to dispose of most of their waste. Of course, the problem in Greece, as elsewhere in much of Southern Europe, was not landfill as such but the very large number of unregulated and ill-managed landfill sites.

As Kemberling and Roberts (this volume) found in Louisiana, it is generally much easier for campaigners to prevent the siting of new waste facilities in or near their communities than it is to secure the closure or removal of already existing facilities. Thus in Greece it was the few proposals for new and better managed landfills that stimulated the most effective protest mobilisations and not the many existing sites about which there were longstanding complaints. One reason for the intensity of struggles over proposed waste facilities is that whereas the neighbours of existing facilities have had time to become resigned to them or perhaps to move away, the proposal of new, and almost invariably much larger facilities raises the prospect of long-term co-existence with an unwanted intrusion as well as fears of the unknown. Botetzagias and Karamichas conclude that while the Greek mobilisations do demonstrate some NIMBY characteristics, such campaigns might more adequately be characterised as *ad hoc* mobilisations reflective of the tensions of late modernity. In Greece, as elsewhere, the public's mistrust of experts and the difference in perceptions of risk between planners and public were stimuli to mobilisations against waste infrastructure.

From local to global?

The story of campaigns against waste infrastructure, especially against waste incinerators, can be told as one of the progression from isolated local struggles through national networking to a transnational, even possibly global, movement. But the story thus told is of a picture painted with a very broad brush and one rather more focused upon the Western hemisphere and the Pacific region than a truly universal one. The environmental justice movement is primarily a US phenomenon, and whilst it has resonances elsewhere, it has not simply been replicated across the globe.

Nevertheless, in recent decades, the links between environmental and social justice issues have been increasingly made. Northern-originated environmental movement organisations have played an important role in this transnational networking. Friends of the Earth is perhaps the most prominent among them, and Friends of the Earth International (FoEI) has grown into a nearly global network of some 80 organisations. However, FoEI is a network of autonomous national organisations and not simply the extension to the global South of a Northern organisation. Moreover, the transmission of ideas and the processes of learning within FoEI does not simply travel from North to South, but is a multi-directional process that has changed the perceptions and campaigning priorities of Northern members at least as much as it has those of Southern affiliates. Thus FoEI has functioned as a conduit for the flow of Southern perspectives on environmental justice to the campaigning organisations of the global North (Doherty 2006, Rootes 2006, Carmin and Bast 2009).

In fact, environmental movements in the global South have long been bound up with issues of democracy and social justice (Haynes 1999), and the resistance of communities in the global South to the imposition of waste incinerators and other such facilities should not be seen as an effect of diffusion from the US (cf. Pellow 2007, pp. 78–79).

We shall in this volume touch only lightly on the transnational dimensions of the waste problem but, as Khoo and Rau (this volume) argue, the transboundary trade in waste is rooted in inequality. Countries in need of foreign exchange or anxious to develop industrially are induced to accept the location on their territory of waste treatment facilities that would be difficult or impossible to site in the countries where that waste was generated. The fact that governments in the host countries are often authoritarian, intolerant of protests and sparing in the information they disclose to citizens makes it much easier to locate hazardous waste treatment facilities there than in fully democratic states.

The increasing effectiveness of resistance to the siting of waste facilities perceived to be risky in democratic states has tended to propel waste offshore. The international trade in hazardous waste is regulated by the Stockholm Convention, but the export of non-toxic waste is not illegal, and a great deal of the material collected for recycling in affluent countries of the global North is shipped for reprocessing in places such as China, India and Brazil. Waste is thus a part of the global flows of goods and services that have greatly increased in recent decades. Resistance to the flow of waste to the global South has, however, increased in the recipient countries, often but not always with the assistance of Northern environmental and human rights NGOs (see, e.g., Pellow 2007).

A particularly interesting development is the formation of GAIA – the Global Alliance for Incinerator Alternatives/Global Anti-Incinerator Alliance – an international network that supports community-based movements for environmental justice. Founded in December 2000 at a meeting in South Africa that attracted people from 23 countries, GAIA grew out a successful campaign in the Philippines to secure a nationwide ban on incineration. It has since grown to be a worldwide network numbering more than 500 member organisations in over 80 countries, with offices in the Philippines, California and Argentina. As well as campaigning against the trade in toxic waste, its major campaign in 2009 was 'Zero Waste for Zero Warming' (http://www.zerowarming.org) with the message that aiming for zero waste – eliminating waste, and increasing reuse, recycling, and composting – addresses the root causes of global warming, safeguarding human health and dramatically reducing demand on natural resources while resisting the efforts of waste companies to sell incinerators as 'renewable energy facilities' (http://www.no-burn.org, accessed 16 September 2009).

GAIA celebrates the successes of campaigners in the Philippines, Malaysia, Indonesia, South Africa and Mauritius in resisting hazardous waste incinerators, but with active campaigns against waste incineration in North and South America and Europe, as well as South-east Asia, and with overlapping affiliations with Greenpeace in Asia and FoE in England, it is clear that campaigning against waste infrastructure is unprecedentedly well networked on a transnational basis (cf. Pellow 2007, ch. 3). Most remarkably, it is a Southern-originated network that has inspired efforts in the North.[2]

Prospects

Changing technologies of waste management

Mass burn incinerators of the kind most commonly employed to burn waste, operating even at high temperatures, are not particularly efficient in doing the one thing that is their most widely touted advantage – reducing the amount of waste consigned to landfill. Depending upon the characteristics of the waste so burned, the volume and weight of residual ash can amount to as much as 30% of that of the waste incinerated. Some of the bottom ash may, with or without further treatment, be suitable for use in construction materials, but the rest must itself be landfilled, and the more heavily contaminated fly ash collected from the emission stacks must be treated as hazardous waste, stabilised, and landfilled in specially controlled landfill sites. More sophisticated fluidised bed incinerators, in which the incoming waste is bombarded with sand in order to increase the efficiency of combustion, produce less residual bottom ash, much of which is suitable for use in construction aggregates, but much greater amounts of toxic fly ash.

Especially as concerns about climate change increase, there is ever greater scrutiny of the efficiency with which various waste technologies contribute to the twin goals of energy output and, directly or indirectly, reduction of emissions of greenhouse gases into the atmosphere. Since most of the energy produced by incineration is heat that can only imperfectly be harnessed to produce steam to drive electricity-generating turbines, incinerators most efficiently contribute to these twin policy goals when the residual heat is captured and distributed to neighbouring buildings. For this reason, incineration with heat recovery (combined heat and power – CHP) ranks above incineration without heat recovery in the efficiency tables (Hogg *et al.* 2008). However, for CHP to be economic, incinerators need to be located very close to potential markets for their heat. Yet the difficulty of finding suitable sites in urban areas, and the general perception of incinerators as bad neighbours, obliges waste authorities to seek sites remote from those most able and likely to object, which in practice often means rural sites remote from any potential market for the incinerator's heat. This, and the large capacities necessary to achieve the economies of scale that make high-tech incinerators financially viable,[3] entails large numbers of vehicle movements that are themselves a major source of pollution and of nuisance.

It is such considerations that have given urgency to the search for waste management practices and treatment technologies that avoid the pitfalls of landfill and of incineration.

No waste treatment facility is ever likely to be welcomed as a neighbour, and local objectors to bio-digesters and composting plants have raised concerns about odour and bio-aerosols, but the fact that bio-digesters can be economic at relatively smaller scale than state-of-the-art incinerators is likely to recommend them to communities concerned by the increased vehicle movements associated with large scale facilities. Because smaller scale waste treatment can be sited closer to the source of waste arisings, it is more consistent with the proximity principle that is an important element of planning for sustainability. And because it may be a source of useful amounts of energy, potentially including even road fuel or direct connection to the national gas grid, it may also make a contribution to the resilience of communities at a time when concerns about energy security are increasing.

Most new waste treatment technologies are new only in the sense that they are newly applied to waste or to municipal solid waste in particular. Nevertheless, the variable contents and character of household waste present obstacles to the simple application of a technology proven in one context with one type of waste to another context and a different type of waste. Thus fluidised bed incineration has been shown to work well with a homogeneous waste stream such as lumber industry waste in Scandinavia, but less well with more mixed and variable municipal waste in the United States and France. Pyrolosis or gasification may work well under laboratory conditions, but used to treat household waste in Australia, it failed to keep emissions of arsenic within permitted levels, probably because of the amount of treated timber that found its way into the waste stream. Thus what works, or is claimed to work, in Japan may not work, or work equally well, in the United States, Britain or Ireland.

Although none of these problems is likely to be technically insoluble, awareness or suspicion of them is enough to deter adoption of new technologies since each new plant may be considered an expensive experiment, especially by waste companies whose engineers are familiar with an older technology proven to work in local conditions. The rationalisation of the waste industry, dominated in Europe as in the US by a shrinking number of ever larger companies, may produce corporations of a scale sufficient to deploy the financial resources to entertain such experiments, but it may also, at least in the short term, militate against innovation. Thus the adoption of newer waste treatment technologies will very often depend upon the guidance and incentive structures established by central governments and/or upon the specifications required by the waste disposal authority.

Perhaps the most compelling policy imperative impacting upon waste in the present century is that of mitigating climate change. For this reason, it is likely that there will be some convergence between US and EU waste management policy because, even with methane capture, landfill, upon which the US heavily depends and which the EU is determined to reduce drastically, is a much greater producer of greenhouse gases than any of the new waste treatment technologies. With the climate so high on policy agendas, incineration, already low on the waste hierarchy on both sides of the Atlantic, may very well be displaced altogether by a variety of technologies better able to deliver more nearly optimal combinations of waste reduction, greenhouse gas minimisation, and recovery of materials and/or energy.

From policy to politics

Facilities employing even the most appropriate waste treatment technology may, however, be resisted if the process by which they are proposed and implemented leads host communities to feel that they have not been adequately and honestly informed about all matters relevant to the technology and its routine operation, including costs as well as benefits, and/or if they feel excluded from the decision-making process or that that process is not transparent. A recurrent theme in the contributions assembled here is people's resentment of and resistance to waste management schemes that are imposed upon them.

Inadequacies of democratic procedures are not peculiar to countries with demonstrably authoritarian regimes. In Greece, anti-landfill campaigners complained about the lack of transparency and democratic accountability in siting decisions, and in Ireland it was the loss of local control over waste management that so diminished trust in the authorities and heightened opposition to proposed waste incinerators (Leonard, Doran and Fagan, this volume). The distribution of waste dumps in the US was in part explained by the structuring of county boundaries, and the high proportion of non-citizens among Hispanics in California, for example, meant that decisions there were taken by legislators representing people other than those most directly affected. So too were decisions on incinerator siting in the English counties, where issues of commercial confidence also obscured waste management decisions from public scrutiny (Rootes, this volume).

Yet while deliberative and inclusionary planning procedures that permit members of potentially affected communities to participate in the decisions that affect them may diminish the democratic deficit of which so many anti-incinerator campaigners complain, they are not a panacea. The modern management of waste, even with newer modular technologies, requires a measure of centralisation of waste treatment such that the few will inevitably play host to the unlovely facilities that treat the waste of the many. A measure of inequity is inescapable and, for that reason, it is likely that the management of waste will continue to be a matter of local contention and to sustain community campaigns for less risky forms of waste management.

Whether such community campaigns will stimulate or sustain social movements is less certain. Present indications are, however, that community campaigns against waste incineration are now unprecedentedly well networked, nationally and transnationally. They are also increasingly well aligned and integrated with the programmatic concern with climate change that now dominates the agendas of national and transnational environmental NGOs. As a result, grassroots campaigns against waste incinerators can, in many countries and at the transnational level, now be considered as strands of a broader environmental movement. Moreover, by connecting the grand narrative of climate change with practical local concerns, they may make an important contribution to the vitality of those environmental movements.

Notes

1. There are, in most countries, relatively few sites that have the geology appropriate for landfill and they are very unevenly distributed geographically and not usually in close proximity to the source of the largest quantities of waste arisings. Some of the largest landfill sites in the US have, it has only gradually become apparent, been sited over geological faults with resultant contamination of groundwater and deep aquifers (see e.g. Luton 1996, on the case of Spokane). Pollution by landfills of groundwater and rivers in the Rhine valley was a major driver of Dutch and German and, ultimately, European Union policy requiring the diversion of waste from landfill.
2. At a meeting in London in May 2006 designed to bring together and encourage networking among British anti-incinerator campaigns, GAIA's Filipino video was playing in the background throughout and the success of its campaign in achieving a nationwide ban on incineration was repeatedly referred to as an example of what could be achieved. The network proposed at that meeting was stillborn, but, with help from FoE, the process to establish the British national anti-incinerator network, UK without incineration network (UKWIN), began later in the year.
3. The optimal size of an energy from-waste/waste-to-energy plant burning MSW is estimated to be 400,000 tonnes per annum. Such plants are capital intensive and, economically to provide CHP, require year-round demand for high pressure steam for heating/cooling (Ilex Consulting 2005).

References

Carmin, J. and Bast, E., 2009. Cross-movement activism: a cognitive perspective on the global justice activities of US environmental NGOs. *Environmental Politics*, 18 (3), 351–370.

Cassee, F., 2009. Health effects of particulate matter. Presentation to Netherlands Environmental Assessment Agency Workshop on 'Measurements and Modelling of $PM_{2.5}$ in Europe', Bilthoven, The Netherlands, 23–24 April.

Chen, H., Goldberg, M.S., and Villeneuve, P.J., 2008. A systematic review of the relation between long-term exposure to ambient air pollution and chronic diseases. *Reviews on Environmental Health*, 23 (4), 243–297.

Chlorine Chemistry Division of the American Chemistry Council, 2003. Backyard trash burning: the wrong answer. Available from http://dioxinfacts.org/sources_trends/trash_burning.html [Accessed 12 September 2009].

Committee on the Medical Effects of Air Pollutants (COMEAP), 2009. *Long-term exposure to air pollution: effect on mortality*, Final Report. London: Health Protection Agency.

Diani, M., 1992. The concept of social movement. *Sociological Review*, 40 (1), 1–25.

Doherty, B., 2006. Friends of the Earth International: negotiating a transnational identity. *Environmental Politics*, 15 (5), 860–880.

European Environment Agency (EEA), 2009. *European Community emission inventory report 1990–2007*. Technical report 8/2009. Copenhagen: EEA.

Garavan, M., 2004. *The patterns of Irish environmental activism*. Thesis (PhD). Department of Political Science and Sociology, National University of Ireland, Galway.

Haynes, J., 1999. Power, politics and environmental movements in the Third World. *Environmental Politics*, 8 (1), 222–242.

Health Protection Authority (HPA), 2009. *The impact on health of emissions to air from municipal waste incinerators*. London: Health Protection Authority.

Hogg, D., *et al.*, 2008. *Greenhouse Gas Balances of Waste: Management Scenarios*. Report for the Greater London Authority. Bristol: Eunomia Research & Consulting Ltd.

Ilex Consulting, 2005. *Extending ROC to energy from waste with CHP*, Supplementary report to DTI. Oxford: Ilex Energy Consulting.

Kousis, M., 2003. Greece. *In*: C. Rootes, ed. *Environmental protest in Western Europe*. Oxford and New York: Oxford University Press, 109–134.

Kousis, M., 2007. Local environmental protest in Greece, 1974–94: exploring the political dimension. *Environmental Politics*, 16 (5), 785–804.

Krohn, W. and van den Daele, W., 1998. Science as an agent of change: finalization and experimental implementation. *Social Science Information*, 37, 191–222.

Laden, F., *et al.*, 2006. Reduction in fine particulate air pollution and mortality. *American J. of Respiratory Critical Care Medicine*, 173, 667–672.

Luton, L.S., 1996. *The politics of garbage: a community perspective on solid waste policy making*. University of Pittsburgh Press.

Miller, K.A., *et al.*, 2007. Long-term exposure to air pollution and incidence of cardiovascular events in women. *New England J. of Medicine*, 356, 447–458.

National Research Council/National Academies of Science (NRC), 2000. *Waste incineration and public health*. Washington, DC: National Academy Press.

O'Brien, M., 2007. *A crisis of waste? Understanding the rubbish society*. London: Routledge.

Pellow, D.N., 2007. *Resisting global toxics: transnational movements for environmental justice*. Cambridge, MA: MIT Press.

Pope, C.A. and Dockery, D.W., 2006. Health effects of fine particulate air pollution: lines that connect. *J. of the Air and Waste Management Association*, 56, 709–742.

Psomopoulos, C.S., Bourka, A., and Themelis, N.J., 2009. Waste-to-energy: a review of the status and benefits in USA. *Waste Management*, 29, 1718–1724.

Rootes, C., 2003. Britain. *In*: C. Rootes, ed. *Environmental protest in Western Europe*. Oxford and New York: Oxford University Press, 20–58.

Rootes, C., 2004. Environmental movements. *In*: D.A. Snow, S.A. Soule, and H. Kriesi, eds. *The Blackwell companion to social movements*. Oxford and Malden, MA: Blackwell, 608–640.

Rootes, C., 2006. Facing South? British environmental movement organisations and the challenge of globalisation. *Environmental Politics*, 15 (5), 768–786.

Saffron, L., Giusti, L., and Pheby, D., 2003. The human health impact of waste management practices: a review of the literature and an evaluation of the evidence. *Management of Environmental Quality*, 14 (2), 191–213.

Schlüter, A., Phillimore, P., and Moffatt, S., 2004. Enough is enough: emerging 'self-help' environmentalism in a petrochemical town. *Environmental Politics*, 13 (4), 715–733.

Shevory, T., 2007. *Toxic burn: the grassroots struggle against the WTI incinerator*. Minneapolis: University of Minnesota Press.

Strategy Unit, 2002. *Waste not, want not: a strategy for tackling the waste problem in England*. London: Cabinet Office.

Tammemagi, H., 1999. *The waste crisis: landfills, incinerators, and the search for a sustainable future*. Oxford University Press.

Environmental movements and campaigns against waste infrastructure in the United States

Christopher Rootes[a] and Liam Leonard[b]

[a]Centre for the Study of Social and Political Movements, School of Social Policy, Sociology and Social Research, University of Kent, Canterbury, England; [b]School of Business and Humanities, Institute of Technology, Sligo, Ireland

Campaigns against waste infrastructure in the US emerged in the 1970s against a background of increasing public anxiety about the impacts of high-tech industrialism upon the environment and human health. Independently of major environmental NGOs, and unlike earlier anti-nuclear campaigns, which also involved grassroots protests, waste campaigners quickly became networked and raised new issues of environmental justice. Initially focused upon landfills and hazardous waste, the environmental justice movement took up and amplified local protests against waste incineration. Independently of popular protest, changes in public policy and the economics of the waste industry also contributed to the unpopularity of waste incineration, and recycling regained appeal. Campaigns against waste infrastructure have contributed to the broadening of the US environmental movement as well as to ecological modernisation.

The United States has long been a participatory society, even perhaps a 'movement society' (Meyer and Tarrow 1997). The vitality and variety of the succession of social movements in the US during the latter part of the twentieth century has made it a virtual laboratory for social movement scholars. This is no less true of contention over environmental issues. Indeed, such is the richness of popular mobilisations over environmental issues in the US that several particular strands of such contention have been referred to as movements in their own right. It is beyond our purpose here to settle the question about whether there is one environmental movement in the US or several. For our purposes it is sufficient to observe that each successive wave of mobilisation has established organisations and broadcast examples of strategic practice and tactical repertoire that have changed the pattern of political opportunities for and influenced the actions of its successors. We focus our attention here upon those dimensions of environmental contention that most directly bear on our concern with struggles over waste and waste infrastructure.

The emergence of the environmental movement

The transformation of the American environment in the wake of European settlement quickly raised questions about the relationship between humankind and nature, questions that became more acute as the pace of settlement increased. By the nineteenth century it was not merely the cutting of forests and ploughing of native grasslands that alarmed some, but the impacts of increasing industrialisation and urbanisation. Seeing the landscapes of pre-conquest America disappearing, preservationists became concerned to protect what was left, especially as westward expansion of the United States brought the dramatic landscapes of the west within its boundaries. Conservationists were alarmed that in the haste to clear land for cattle and farming, forest resources were being squandered without proper consideration of future needs. Before the end of the nineteenth century, organisations promoting the preservation of the natural environment and the conservation of natural resources began to proliferate, and in the twentieth century they came to constitute an important advocacy community in Washington (see, e.g., Brulle 2000, Bosso 2005). They did not, however, constitute a grassroots social movement, and their increasing preoccupation with lobbying in Washington rather than responding to grassroots concerns was to become a source of increasing dissatisfaction, leading in time to the rise of new organisations and networks.

The increasing pace of industrialisation after World War II visibly increased the pressure upon the natural environment just at the time that rising affluence increased people's opportunities to appreciate it. Major infrastructure projects proposed to serve the needs of a growing urban population began to encounter greater and more concerted opposition than in the past. In the 1950s, conservation organisations successfully resisted the construction of a dam at Echo Park in the Dinosaur National Monument, but failed to prevent construction of the Glen Canyon dam on the Colorado River, and that failure rankled (Shabecoff 2003, p. 149). In the early 1960s it was an ad hoc local group that successfully petitioned the courts to block plans to build a hydro-electric pumping station on Storm King mountain in New York state. The court's ruling in that case that nature was deserving of legal protection is credited with laying the foundations for environmentalists' unprecedentedly frequent recourse to litigation in pursuit of their aims (Shabecoff 2003, pp. 95–96).

Nature may have found its defenders, but the technological developments associated with continuing industrialisation carried with them environmental impacts of a novel and more pervasive kind.

The event most often credited with kick-starting the modern environmental movement in the US was the publication in 1962 of Rachel Carson's book *A Silent Spring*, which highlighted concerns about the impact of science on nature. Carson's earlier books on marine life were already best-sellers and, at a time when the US Department of Agriculture's programme of compulsory insecticide spraying had provoked widespread controversy and litigation, the publication of *A Silent Spring* was carefully calculated to have maximum impact. First serialised in the *New Yorker*, then a Book-of-the-Month Club selection, the book, the extensive media coverage and the controversy it ignited, highlighted for the general public the impact of the indiscriminate use of synthetic pesticides such as DDT on wildlife in the years following the introduction of scientised methods of pest control in agriculture. Public concern and scientific debate about Carson's work led President John F. Kennedy to call on the Science Advisory Committee to investigate issues surrounding the use of pesticides. That inquiry confirmed Carson's contention, led to tighter regulation of the use of chemical pesticides, and is widely credited with having laid the foundations for the formation of the Environmental Protection Agency in 1970.

Unease about the environmental impacts of industrialism, compounded by rising opposition, especially among students, to the US prosecution of the war in Vietnam, contributed to the emergence of a countercultural movement that was sharply critical of the subordination of nature by humankind in the industrial era and echoed the Romantic rejection of industrialism and instrumental rationality of the late eighteenth and early nineteenth centuries. Towards the end of the decade, environmental concern was also increased by highly publicised mishaps such as the 1969 fire on the Cuyahoga River and the oil spill off the California coast at Santa Barbara.

Heightened awareness of such a range of environmental issues spread among politicians and the general public. The first Earth Day, 22 April 1970, organised in just a few months and tapping into a growing undercurrent of environmental concern, attracted some 20 million participants (Graham 1999, p. 2). A flurry of legislation ensued, to establish the Environmental Protection Agency (EPA) and to protect air, water, coasts and public lands. Rising public concern and dissatisfaction with the established preservationist and conservationist organisations stimulated the formation of a new wave of environmental organisations: Friends of the Earth was founded in 1969, the Natural Resources Defense Council (NRDC) in 1970, and Greenpeace in 1971. These and other new organisations would be vehicles and resources for environmental campaigners in the years that followed.

The oil crisis of 1973 and the ensuing economic recession quickly sank environmental issues down the order of public concerns and raised energy security close to the top of the political agenda. The result was renewed interest in the exploitation of nuclear fission to produce electricity and a corresponding increase in local campaigns against proposed nuclear power plants.

Anti-nuclear campaigns

Controversies over the siting of nuclear facilities were not new. Objectors had successfully opposed the construction of nuclear power stations at Bodega Bay, California, in 1958 and later at Malibu, principally on environmental grounds. These local campaigns attracted support from the Sierra Club, and are regarded as the start of the US anti-nuclear movement (Wellock 1998). They and the campaign against the nuclear power plant proposed for Diablo Canyon, California, a decade later included a strong strand of environmental concern and of a view of nature at odds with that of industrial society (Wills 2006). Gradually, concerns about the direct environmental impacts of nuclear power plants, fears about possible radioactive emissions to air and water, and worries about reactor safety coalesced just at the time when the US government, concerned to ensure the nation's energy security, was encouraging the construction of large numbers of nuclear power stations to reduce dependence on increasingly imported oil.

Differences over whether and how to oppose the siting of a nuclear power station at Diablo Canyon caused divisions in the Sierra Club in the 1960s that crystallised the emergence of ecological concerns distinct from those of earlier conservationism, but Epstein (1991, p. 58) dates the start of the non-violent direct action movement of the late 1970s and 1980s to the formation of the Clamshell Alliance at Seabrook, New Hampshire, in 1976. The first major anti-nuclear protests in the US were at Seabrook nuclear power plant in 1977 and 1978. In 1978, 500 people were arrested at a protest at Diablo Canyon. Anti-nuclear campaigners continued to grow in numbers and impact through the 1970s, and succeeded in stopping or slowing the development of many reactors, but it was the partial meltdown of the reactor core at the Three Mile Island nuclear plant at Harrisburg, Pennsylvania, in March 1979 that delivered the coup de grace. In May 1979, 70,000 people attended a 'No Nukes' rally against nuclear power in Washington, DC, followed by a rally of 200,000 people in New York later in the year; nearly 40,000 people attended a protest rally at Diablo Canyon (Gottlieb 2005, p. 240).

1979 may have marked the high point of the popular mobilisation against nuclear power in the US, but local protests continued for more than a decade. Some of the nuclear plants that provoked such protest were eventually built and commissioned, but plans for many more were thwarted. After 1979, plans for more than 40 nuclear power stations were cancelled, two under construction were converted to burn coal or gas, and more than 20 existing nuclear facilities were closed. Three Mile Island so amplified public anxiety about nuclear energy that for a generation expansion of the US nuclear industry was halted in its tracks.

But was there an anti-nuclear *movement*? By the mid-1970s, anti-nuclear protest in the US was no longer limited to local protests but had gained wider support and influence. It may nevertheless be stretching the definition to categorise as a social movement protests that had neither a single co-ordinating organisation nor a uniform set of goals, but the 'movement' of opposition to nuclear power had succeeded in attracting a great deal of national attention (Walker 2004, pp. 10–11). Opposition to nuclear energy might never have been so successful without Three Mile Island, but Three Mile Island may not have had such an impact had not the perceived risks of nuclear energy been so amplified by anti-nuclear campaigners (see Agnone 2007). The catastrophic meltdown at the Chernobyl nuclear plant in the Ukraine in 1986 simply confirmed such fears and solidified public opposition to nuclear power (Rosa and Freudenburg 1993, pp. 48–49).

The public fears of nuclear technology, and the lack of trust in the nuclear industry to prioritise public safety, made the subsequent search for a permanent repository for the US's nuclear waste especially fraught. Public opposition, quickly translated into political opposition by Congressional representatives, led to the rapid elimination of most of the proposed sites (Dunlap *et al.* 1993). By the time that Yucca Mountain, Nevada, was pragmatically identified as the sole remaining possible site it had become politically impossible to take the decision. It remains unresolved, and so nuclear waste remains in temporary storage at power stations, and the inability of the industry to find a permanent safe storage site is a severe constraint upon its possible revival.

The environmental justice movement

It was against this background of such protests against industrial facilities of questionable safety that what came to be known as the environmental justice movement (EJM) emerged. The EJM is an inclusive term used here to embrace both the 'anti-toxics movement' and the 'people of colour environment movement' that emerged a little later (Pellow and Brulle 2005, p. 8). Whilst these strands of environmental activism have somewhat different agendas – the former focusing upon potentially universal complaints about polluting processes and practices and the threats they pose to public health, the latter raising complaints about the inequitable distribution of environmental ills (Cf. Cable *et al.* 2005, pp. 59–60) – both employ a discourse of environmental justice to advance their claims, and they have in practice tended to overlap (Szasz 1994, Schlosberg 1999), even if their convergence is less than some think necessary and desirable (see, e.g., Brulle and Pellow 2005, pp. 298–299).

Although protests by African-Americans have dominated much of the coverage of the EJM, it was events in a mostly white, working-class community in upstate New York that brought the spotlight of publicity to toxic waste. In 1978, residents of Love Canal, a suburb of Niagara Falls, experiencing a high incidence of unexplained illnesses, miscarriages and birth defects, discovered that a primary school and their homes had been built upon and adjacent to a site previously used as a dump for over 20,000 tons of toxic chemical industry waste. This revelation, and the residents' protests, attracted media attention and provoked President Carter to declare a federal health emergency and to order the expenditure of federal emergency funds to relocate affected residents. Residents continued their protests to force effective action, at one point holding EPA officials hostage for five hours. Within two years, most of the 1000 families affected sold their houses to the government and moved away, and the Comprehensive Environmental Response, Compensation, and Liability Act or 'Superfund' Act was signed into law and established a multibillion-dollar programme to clean up hazardous waste sites, of which some 1000 were declared a national priority.

One enduring outcome of the Love Canal fiasco was the Citizens Clearinghouse for Hazardous Wastes (CCHW) (later renamed the Center for Health Environment and Justice, CHEJ), founded in 1981 by Lois Gibbs, the principal organiser of the Love Canal residents. Gibbs was convinced of the importance of grassroots campaigning, and so CCHW aimed to provide information, advice and support to otherwise isolated local communities campaigning against all manner of hazardous wastes. Given its origins, it is not surprising that the Landfill Moratorium Campaign should have been one of its most prominent campaigns. Its success has been such that, while dozens of new hazardous waste landfills have been proposed, in the whole of the US only one has been opened since CCHW began its work (http://www.chej.org/history. htm, accessed 4 September 2009).

The environmental justice movement developed further during the 1980s as communities became more aware of and resistant to the siting of hazardous plants or dumps. Because these facilities were often located, sometimes deliberately (Cole and Foster 2001), in economically disadvantaged non-white neighbourhoods, the environmental justice movement became associated with the civil rights movement. Attempts to dump toxic waste in the primarily African-American community of Afton, Warren County, North Carolina in 1982 led to protests and arrests. In 1987 the Commission for Racial Justice published its *Toxic Wastes and Race in the United States* report, which appeared to confirm the extent of dumping in or near minority communities.[1] Charges of environmental racism and the publication of Robert Bullard's *Dumping in Dixie: Race, Class, and Environmental Quality* (1990) raised the stakes. In 1991, environmental justice was a major concern of the First National People of Color Environmental Leadership Summit in Washington, DC. Official response was swift: in 1992, the Environmental Protection Agency created the Office of Environmental Justice (Shabecoff 2003).

One consequence of the campaign against new hazardous waste landfills has been a series of proposals to extend existing landfills, but it is testimony to the impact of the environmental justice movement that extensions to landfills in the 1990s were *less* likely to have been approved in or near communities whose populations consisted disproportionately of ethnic minorities rather than the white majority (Atlas 2001a).[2]

The environmental justice movement began with protests that were mostly local and unconnected to major environmental organisations. Like later anti-nuclear protests (Epstein 1991), they drew on and celebrated the legacies they inherited from a long line of American popular struggles (Szasz 1994, p. 150) but, unlike the anti-nuclear campaigns, they quite quickly became networked, both by the institutions that had grown from the civil rights movement and, more generally, by networks that grew out of local struggles themselves (Schlosberg 1999). Public awareness of the hazards of improper waste disposal was greatly increased by the flurry of media attention devoted to Love Canal, and that media attention stimulated other new cases and revived other, older campaigns that had never enjoyed media attention. Before 1978, protests against industrial contamination were sporadic and isolated; after 1978, the majority of reported cases were networked with others, and after 1980, with the organisational infrastructure of a movement established, campaigns became both widespread and well connected one to another (Szasz 1994, pp. 70–72).

Freudenberg and Steinsapir (1992, pp. 33–35) list a number of achievements of what they call 'the grassroots environmental movement': they have contributed to improving public health; they have forced corporations to consider the environmental consequences of their actions, either by making continued pollution unprofitable, or by persuading polluting companies that it would be cheaper to clean up their operations than engage in protracted battles with residents; the cumulative effect of successive local struggles created political and economic pressures for preventative measures and, by closing off the polluting options, encouraged cleaner production, safer products and more recycling; they empowered otherwise depressed afflicted communities; they took environmental concerns to working-class and ethnic minority people who had been untouched by established environmentalism; and they helped to reshape public opinion about the environment and public health. Even where they failed to achieve particular objectives, grassroots campaigners won rights of public participation in decision-making and the right to information about potentially polluting projects. Perhaps most importantly, they developed organising skills, enhanced political efficacy and fostered social networks among those involved in campaigns, and schooled 'movement entrepreneurs' who went on to organise or support campaigners on other issues or in other places. Thus one round of grassroots mobilisation fertilised the ground for those that followed.

Campaigns against incineration

One product of increasing awareness of the problems with landfill was a search for more sanitary alternatives, and in a country where a growing population and rising affluence were contributing to greater amounts of waste, the problem was increasingly urgent.

The waste problem was not new. Nor was this the first time that incineration had been seen as a means of its solution. The difficulties of disposing of municipal waste in a sanitary manner had increased as the US became more urbanised, and the urban sanitation 'movement' was a powerful response to the health-threatening filth of US cities at the end of the nineteenth and in the early twentieth centuries (Melosi 2005). Appropriately, soon after the first waste incinerator was constructed in England in 1875, the first waste incinerator in the US was built at Governor's Island, New York in 1885. By 1914, there were some 300 waste incinerators in the US; by 1960 there were about 600 before their numbers began to decline as these ageing incinerators faced increasing concerns about air pollution and competition from suitably engineered sanitary landfills (Walsh *et al.* 1997, p. 2; Melosi 2005, p. 187).

Management of municipal waste was a local responsibility, but inconsistent and often incompetent management of waste disposal facilities, mostly landfills but also incinerators, gradually drove the issue up the political hierarchy and in 1965 the first federal legislation governing the management of solid waste was enacted. It was, therefore, not surprising that the federal government should be concerned to find a solution to the problem of increasing waste arisings or that it should be joined to other pressing policy concerns. The Solid Waste Disposal Act (1965) was followed by the Air Quality Act (1967) and the Resource Recovery Act (1970). The 1965 Act encouraged the development of sanitary landfills, that of 1967 required incinerators to be fitted with scrubbers to remove pollutants, and the 1970 Act encouraged materials recovery and recycling.

In the 1960s and 1970s, activist-led grassroots recycling initiatives had proliferated, most of them started and staffed by volunteers whose motives were community-building as well as environmental protection. However, the formation of the EPA from 1970 channelled environmental protest into negotiation and fostered the development of a range of attendant professionalised lobbying and legal advocacy. Local grassroots organisers and their ideals were marginalised. Moreover, lack of markets for recyclates discouraged recycling efforts, and in the 1974 recession many schemes were abandoned. Although recycling had, alongside waste-to-energy (WtE), been incorporated in the resource recovery frame valorised by the 1970 Resource Recovery Act, because the stronger voices were those of the waste corporations and institutionalised actors, recycling disappeared from the discourse and resource recovery became equated with WtE (Lounsbury *et al.* 2003, pp. 83–84).

These developments conjoined with the energy crisis of 1973–1974 to make a new generation of incinerators capable of burning waste to generate electricity attractive and economically viable, the increased price of oil having tripled the price of electricity, and the federal government obliging by requiring local utilities to buy the electricity so generated. By the late 1970s, these WtE plants also gained from the engineering capacity made redundant by the travails of the nuclear power industry. The Resource Conservation and Recovery Act 1976, by focusing attention upon hazardous wastes and the need to close unsanitary landfills, created opportunities for proponents of both incineration and recycling. So many unsanitary landfills closed that the cost of landfill more than doubled between 1982 and 1988. Suddenly the economics of WtE looked much more attractive (Lounsbury *et al.* 2003, p. 86).

Concerned to manage increasing volumes of waste whilst accommodating public anxieties about landfill, and encouraged by tax incentives, developers and waste planners turned to WtE. The numbers of WtE plants operating, under construction or proposed rocketed from 60 in 1980 to 200 by 1985 (Walsh *et al.* 1997, pp. 6–7). But the boom was short-lived. No new waste incinerators have been constructed since 1996, and there were in 2009 just 88 waste-to-energy plants in operation in 25 states (Psomopoulos *et al.* 2009).

The demise of the waste incinerator in the US has been credited to a remarkably successful national anti-incineration movement in the 1980s, whereafter very few new incinerators were commissioned and many proposed projects failed to gain approval (Walsh *et al.* 1997, p. 244). Many older incinerators closed: of the 186 municipal waste incinerators in 1990, only 112 were still operating in 2002 (Tangri 2003, p.66); of the 6200 medical waste incinerators in 1988, only 115 remained in 2003. Tangri (2003, p. 65) claims that

> the rapid expansion of incineration sparked one of the largest and most effective grassroots environmental movements in American history. In approximately 15 years, this loosely-linked network of mostly volunteer activists succeeded in stopping over 300 proposed municipal waste incinerators across the country, and in imposing increasingly strict air emissions standards, effectively killing off the American municipal waste incinerator industry.

With concerns about the hazardous by-products of industrial processes already heightened by the rash of local campaigns concerning landfills that followed Love Canal, by the mid-1980s, scientists such as Barry Commoner and Paul Connett began to highlight the health risks posed by the emissions released by the process of incineration. Emissions from incineration contained potentially carcinogenic chemical compounds such as dioxins, arsenic and toxic metals such as cadmium but, following Commoner's intervention, it was the dioxin content of incinerator emissions that became the most prominent theme of protests. Soon an anti-incineration movement emerged, centred in the northeast, but with links to existing anti-toxics and environmental groups (Walsh *et al.* 1997).

The environmental justice movement enabled effective networking among otherwise isolated local campaigns. Despite its original primary focus upon hazardous waste and landfills, by 1986 CCHW/CHEJ was an important resource for anti-incineration campaigners; so too, albeit in a smaller way, were the National Toxics Campaign and Greenpeace (Szasz 1994, pp. 72–77, Walsh *et al.* 1997, pp. 157, 170). Concerns about hazardous waste, incineration, dumps and environmental justice coalesced into something that could be represented as a broad anti-toxics movement that, by extending links nationally and ultimately globally, enabled community campaigners to present themselves as more than merely NIMBY ('not in my backyard') protesters; the backyard in question became not merely 'mine' but 'anybody's' and, ultimately, everybody's.

It was not, however, only emissions to air that concerned critics of incineration. Also highlighted by campaigners were the problems posed by incinerator ash. Incineration does not eliminate waste but merely reduces it, depending upon the process, typically to between 15–30% of the weight of the original solid waste material and 5–15% of its volume. A substantial amount of residual ash remains to be disposed of to landfill, but the more successful an incinerator operator is in removing toxins from the air, the more toxic is the material concentrated in that residual ash (Walsh *et al.* 1997, p. 11). Whilst heavy metals may contaminate the bottom ash that comprises the major part of incinerator residues, the fly ash recovered from incinerator emission stacks is especially prone to concentrate toxins and so must be treated as hazardous waste that in turn poses special problems of safe disposal. As part of its work to ensure that dangerous wastes continued to be classified as such, CHEJ in 1988 launched its Kick-Ash Campaign to prevent the classification of solid incinerator ash as waste that might be disposed of without first testing it for toxicity. The result was the defeat of federal legislation that would have permitted incinerator ash, without prior assessment of its toxicity, to be dumped in municipal landfills (http://www.chej.org/history.htm, accessed 4 September 2009). By this time, however, the battle against waste incineration was well advanced.

Another strand of the opposition to waste incineration came from the advocates of recycling. In 1978, the National Recycling Coalition (NRC) was formed with the aim of linking communities, corporations and policymakers interested in recycling to enable them to enter the mainstream debate on solid waste. The NRC promoted recycling on a for-profit basis, with promotion of kerbside collection by mainstream waste management companies and authorities seen as a service to householders rather than a community-building exercise. State- and community-level recycling organisations also sought to change policy and practice, usually in support of non-profit models, but while some, as in California, pre-dated the NRC, others were established under its auspices. The Institute for Local Self Reliance (ILSR), established in 1974, became a key node in the networking of grassroots recycling groups and in the 1980s brought dynamism to NRC.

This coalition came apart in the mid-1980s as NRC attempted to maintain its openness to the mainstream waste industry by emphasising that it saw recycling as a complement rather than an alternative to WtE. This ILSR rejected both because it neglected community-building and grassroots recycling, and because ILSR saw WtE as a threat both to the environment, because of its emissions to air and production of ash, and to continuation of recycling with which it would compete for materials (Lounsbury *et al.* 2003, pp. 88–89). Thus another set of networks was engaged in the struggle against waste incineration.

By the end of the 1980s, the resource recovery frame was decomposed, and recycling was clearly decoupled from WtE and had emerged as the more prevalent discourse. The for-profit recycling industry grew rapidly from 1988, and in 1989 the EPA declared that recycling had an important role to play as an alternative to landfill and thereafter promoted it aggressively. The solid waste corporations, for their part, came to see recycling as an extension of their waste management activities and, since they did not have expertise in waste incineration, they became less and less supportive of WtE (Lounsbury *et al.* 2003, pp. 91–93).

The economics and politics of waste

Anti-incinerator campaigners and their allies in the wider environmental justice movement undoubtedly raised the political temperature surrounding waste management issues, informing and inflaming public opinion against them, and so affected decisions about whether and where to build waste incinerators locally. But there were other forces at work.

The Clean Air Act of 1990, a direct legacy of the environmental protection legislation of the 1970s, mandated 'maximum achievable control technology' to minimise toxic emissions to air. Although the Act was formulated and passed against a background of mushrooming membership of environmental organisations and increasing public concern about the environment (partly in reaction against the anti-environmental efforts of the preceding Reagan administration), it was a product of initiatives by President George Bush, Senior, reinforced by more ambitious proposals emanating from Congress, rather than direct popular mobilisation (Bailey 1998). A knock-on effect, was that between 1990 and 2004 tax credits for plants producing electricity from waste were rescinded.

Although growing scepticism about landfills and the energy crisis of the 1970s had led government to encourage WtE, as energy prices settled, its economic feasibility became a chronic problem. Few WtE plants ever produced sufficient power to make them viable, and their economics worsened as tighter emissions controls increased costs. Moreover, with the increase in the number of large, inexpensive regional sanitary landfills and the relatively low price of electricity, incinerators were not able to compete for their 'fuel', i.e., waste (Melosi 2005, p. 218).

Their difficulties have been compounded by the success of recycling, encouraged by 'zero waste' campaigns and reinforced by EPA targets. Per capita generation of municipal solid waste peaked at 4.65 pounds (2.1 kg) per day in 2000 and has since remained relatively steady. The per capita weight of waste sent to landfills after recycling, composting, and energy recovery has remained virtually static at 2.5 pounds (1.14 kg) per day since 1960. Thus virtually the entire increase in waste has been treated in ways other than landfilling (EPA 2008).

Renewed concerns about energy security in the present century have produced a new flicker of interest in waste as a potential source of energy as an alternative to combustion of fossil fuels. However, despite increased energy prices, the prospects for a renaissance of waste-to-energy in the US do not look promising. Incineration continues to be beset by the obstacles of local concerns about emissions, perceived competition with recycling, siting and financing problems, and low levels of federal government support. Changes in federal policy might, however, change the economic calculation. The Renewable Energy Production Incentive (REPI), created by the Energy Policy Act of 1992, offers tax credits to incentivise the production of renewable energy and, in its most recent reauthorisation (2009), the incineration of municipal solid waste, which did not previously qualify, was reclassified as a source of renewable energy (http://www.epa.gov/osw/hazard/wastemin/minimize/energyrec/rpsinc.htm, accessed 11 August 2009). However the level of federal support thus offered is exceedingly modest: electricity generated by new facilities benefits from a production tax credit of 1 cent per kWh for a maximum of ten years, less than half the credit given to other renewables, and less even than the federal support for electricity produced by burning coal (Glover and Mattingly 2009, p. 4).

Although this most recent change in US policy may increase the economic attractions of waste-to-energy incineration, it would be surprising if, given the available alternatives and the enduring public suspicion of incineration, the result were to be a large-scale programme of incinerator building. This is especially so in light of the increasing availability of new waste treatment technologies that, because they are more capable of operating efficiently in small plants local to the source of waste, and so reduce the requirement to truck waste from one area to another, are likely to be more acceptable to local communities and to environmental policymakers alike (see Rootes, this volume).

Toward zero waste?

Recycling gained so much support because the activist advocates of recycling succeeded in drawing public attention to the drawbacks of incineration, and in making common cause with dominant players in the waste industry. Although not-for-profit recyclers were uneasy about NRC's links with for-profit waste corporations, most remained affiliated with NRC. However, in 1995 they established their own umbrella organisation, the Grassroots Recycling Coalition, which nominally works with NRC but exists mainly to promote its members' own policies and to represent their interests. The irony is that the efforts of early community-based recyclers should have laid the foundations for the commercial recycling operations that now dominate, but recycling is now big business and the scope for revival of community-based not-for-profit recycling appears limited (Lounsbury *et al.* 2003, pp. 94–95).

Campaigners against incineration, Paul Connett prominent among them, have increasingly promoted recycling as an alternative and adopted zero waste as an objective. In 1998 Seattle, long a national leader in environmental matters, adopted a sustainability strategy with zero waste as its ultimate goal. Seattle was quickly followed by several other west coast cities, and by 2007 was one of several, along with Boulder, Colorado, to have adopted a zero waste strategy. It is unlikely that any of these cities would have done so but for sustained pressure from activists concerned about waste and the environmental costs of landfill and incineration. Nevertheless, the increasingly widespread adoption of zero waste strategies demonstrates the extent to which opposition to older forms of waste management has become institutionalised. Campaigners against waste infrastructure may not have won every battle, but they have succeeded in pushing the US toward more sustainable waste management practices.

Some observers worry that one effect of the strength of resistance to unsound waste treatment within the US has been to propel waste offshore to poorer states in the global South (Brulle and Pellow 2005). Pellow (2007, pp. 121, 135–141) describes much of the recycling effort in the US as a form of 'delayed dumping' because over half of the plastic waste collected from US households ends up in Asia for reprocessing. The international trade in non-toxic waste is not illegal, and it is not clear that it is necessarily environmentally undesirable, but the fact remains that insofar as US householders are unaware of the destination of the waste they so assiduously collect for recycling, the job is at best half done. If environmental justice is to be fully achieved it must be on a global basis. Fortunately, there is evidence that both US environmentalists and the Obama administration are aware of this and it is unlikely that difficulties in siting waste treatment facilities in the US will again result in the exportation of toxic wastes. It is noteworthy that the zero waste ambition has rapidly spread globally and that when US cities justify their own adoption of zero waste strategies they often cite precedents from other parts of the world as well as the US.

Conclusion

Campaigns against waste infrastructure have highlighted neglected dimensions of the damage that industrialisation and rampant consumerism have done to the environment. They have raised questions about environmentally damaging practices, processes and products that had seemed only to confer benefits upon society, and in so doing they have contributed to the raising of environmental standards, and thus to efforts to reduce the threats posed by pollution to human health as well as the natural environment. They have also, by focusing upon waste, contributed to the recognition that waste very often represents a waste of resources, and that environmental sustainability often comes with the side-benefit of economic efficiency (Schlosberg and Rinfret 2008).

They have impacted as well upon the environmental movement itself. Campaigns against waste infrastructure have extended the range, thematically and socially, of the US environmental movement, and so have contributed to its reinvigoration. They have introduced to environmental campaigning the very social groups that suffer most from environmental degradation but that in the past mobilised least against it. As a result, the environmental movement no longer has quite so white a face, and no longer are preservationist and conservationist concerns seen as the prerogative of social elites.

In a period in which the dominant narrative about the US environmental movement is that it has lost its way and become trapped within the Washington beltway (Dowie 1995, Bosso 2005, Brulle and Jenkins 2008), grassroots campaigns and networks of grassroots campaigners have introduced organisational innovation and diversity that has not gone entirely unnoticed by the more established environmental NGOs. Grassroots campaigners have performed a discovery role for those larger NGOs and so have contributed to the broadening of the scope of the issues they address (Carmin 1999; cf. Brick and Cawley 2008). In so doing, they have helped to erode the division between nature-focused and human-centred versions of environmentalism, a division that, in the era of global climate change, appears increasingly untenable.

Challenges remain. Perhaps the foremost is to tackle the greatest waste problem of our time – that of greenhouse gases which are the by-product of human combustion of fossil fuels. There are signs that the US, so recently an unrepentant profligate in the use of such fuels and an unabashed refusenik in the struggle to mitigate climate change, may at last be playing a positive role. That it is doing so is in no small measure a tribute to the efforts of those people who campaigned to put climate change on the national agenda.

Acknowledgements
We thank Robert Brulle and David Schlosberg for comments on earlier drafts.

Notes

1. This and other studies that claimed to present evidence of the disproportionate siting of waste facilities in neighbourhoods populated by ethnic minorities were subsequently challenged on methodological grounds, and their implications have not always been considered benign (see, e.g., Foreman 1998). But see also Allen *et al.* (2001) who concluded that people of colour did indeed suffer disproportionately from environmental ills.
2. In fact, Atlas (2001b) found that, allowing for the seriousness of the risks concerned, the poor and people of colour were *not* more likely to be exposed to possible harm from the transport, processing or disposal of hazardous wastes; the large facilities which treated the largest amounts of the most hazardous waste were, if anything, more often close to white than non-white populations.

References

Allen, D.W., Hill, K.M., and Lester, J.P., 2001. *Environmental injustice in the United States: myths and realities*. Boulder, CO: Westview.

Angone, J., 2007. Amplifying public opinion: the policy impact of the US environmental movement. *Social Forces*, 85 (4), 1593–1620.

Atlas, M.K., 2001a. Testing for environmental racism again: an empirical analysis of hazardous waste management capacity expansion decisions. Paper presented at the Annual Meeting of the American Political Science Association, August, San Francisco.

Atlas, M.K., 2001b. Safe and sorry: risk, environmental equity, and hazardous waste management facilities. *Risk Analysis*, 21 (5), 939–954.

Bailey, C.J., 1998. *Congress and air pollution*. Manchester University Press.

Bosso, C., 2005. *Environment Inc.: from grassroots to beltway*. Lawrence: University Press of Kansas.

Brick, P. and Cawley, R.McG., 2008. Producing political climate change: the hidden life of US environmentalism. *Environmental Politics*, 17 (2), 200–218.

Brulle, R.J., 2000. *Agency, democracy and nature*. Cambridge, MA: MIT Press.

Brulle, R.J. and Jenkins, J.C., 2008. Fixing the bungled US environmental movement. *Contexts*, 7 (2), 14–18.

Brulle, R.J. and Pellow, D.N., 2005. The future of environmental justice movements. *In*: D.N. Pellow and R.J. Brulle, eds. *Power, justice and the environment: a critical appraisal of the environmental justice movement*. Cambridge, MA: MIT Press, 293–300.

Bullard, R., 1990. *Dumping in Dixie: race, class, and environmental quality*. Boulder, CO: Westview Press.

Cable, S., Mix, T., and Hastings, D., 2005. Mission impossible? Environmental justice activists' interactions with professional environmentalists and with academics. *In*: D.N. Pellow and R.J. Brulle, eds. *Power, justice and the environment: a critical appraisal of the environmental justice movement*. Cambridge, MA: MIT Press, 55–76.

Carmin, J., 1999. Voluntary associations, professional organisations and the environmental movement in the United States. *Environmental Politics*, 8 (1), 101–121.

Carson, R., 1962. *A silent spring*. New York and London: Houghton Mifflin.

Cole, L.W. and Foster, S.R., 2001. *From the ground up: environmental racism and the rise of the environmental justice movement*. New York and London: New York University Press.

Commission for Racial Justice, 1987. *Toxic Wastes and Race in the United States*. New York: United Church of Christ.

Dowie, M., 1995. *Losing ground: American environmentalism at the close of the twentieth century*. Cambridge, MA: MIT Press.

Dunlap, R., Kraft, M., and Rosa, E., eds., 1993. *Public reactions to nuclear waste: citizens' views of repository siting*. Durham, NC: Duke UP.

EPA (Environmental Protection Agency), 2008. Municipal solid waste generation, recycling, and disposal in the United States: facts and figures for 2007. Available at http://www.epa.gov/osw/nonhaz/municipal/pubs/msw07-fs.pdf [Accessed August 2009].

Epstein, B., 1991. *Political protest and cultural revolution: nonviolent direct action in the 1970s and 1980s*. Berkeley: University of California Press.

Foreman, C., 1998. *The promise and peril of environmental justice*. Washington, DC: Brookings.

Freudenberg, N. and Steinsapir, C., 1992. Not in our backyards: the grassroots environmental movement. *In*: R.E. Dunlap and A.G. Mertig, eds. *American environmentalism: the US environmental movement, 1970–1990*. New York: Taylor and Francis, 27–37.

Glover, B. and Mattingly, J., 2009. Reconsidering Municipal Solid Waste as a Renewable Energy Feedstock, EESI Issue Brief. Washington, DC: Environmental and Energy Study Institute. Available at www.eesi.org [Accessed August 2009].

Gottlieb, R., 2005. *Forcing the spring: the transformation of the American environmental movement*. Washington, DC: Island Press.

Graham, M., 1999. *The morning after Earth Day*. Washington, DC: Brookings.

Lounsbury, M., Ventresca, M., and Hirsch, P.M., 2003. Social movements, field frames and industry emergence: a cultural–political perspective on US recycling. *Socio-economic Review*, 1, 71–104.

Melosi, M., 2005. *Garbage in the cities: refuse reform and the environment*. Pittsburgh, PA: University of Pittsburgh Press.

Meyer, D.R. and Tarrow, S., 1997. *The social movement society: comparative perspectives*. Lanham, MD: Rowman and Littlefield.

Pellow, D.N., 2007. *Resisting global toxics: transnational movements for environmental justice*. Cambridge, MA: MIT Press.

Pellow, D.N. and Brulle, R.J., eds., 2005.*Power, justice and the environment: a critical appraisal of the environmental justice movement*. Cambridge, MA: MIT Press.

Psomopoulos, C.S., Bourka, A., and Themelis, N.J., 2009. Waste-to-energy: A review of the status and benefits in USA. *Waste Management*, 29, 1718–1724.

Rosa, E. and Freudenburg, W., 1993. The historical development of public reactions to nuclear power. *In*: R. Dunlap, M. Kraft, and E. Rosa, eds. *Public reactions to nuclear waste: citizens' views of repository siting*. Durham, NC: Duke UP, 32–63.

Schlosberg, D., 1999. *Environmental justice and the new pluralism*. Oxford and New York: Oxford University Press.

Schlosberg, D. and Rinfret, S., 2008. Ecological modernisation, American style. *Environmental Politics*, 17 (2), 254–275.

Shabecoff, P., 2003. *A fierce green fire: the American environmental movement*. Washington, DC: Island Press.

Szasz, A., 1994. *Ecopopulism: toxic waste and the Movement for Environmental Justice*. Minneapolis: University of Minnesota Press.

Tangri, N., 2003. *Waste incineration: a dying technology*. Quezon City and Berkley, CA: GAIA.

Walker, J.S., 2004. *Three Mile Island: a nuclear crisis in historical perspective*. Berkeley, CA: University of California Press.

Walsh, E., Warland, R., and Clayton Smith, D., 1997. *Don't burn it here: grassroots challenges to trash incineration*. University Park, PA: Penn State University Press.

Wellock, T.R., 1998. *Critical masses: opposition to nuclear power in California, 1958–78*. Madison: University of Wisconsin Press.

Wills, J., 2006. *Conservation fallout: nuclear protest at Diablo Canyon*. Reno: University of Nevada Press.

When time is on their side: determinants of outcomes in new siting and existing contamination cases in Louisiana

Melissa Kemberling[a] and J. Timmons Roberts[b]

[a]Alaska Native Tribal Health Consortium, Anchorage, Alaska, USA; [b]Center for Environmental Studies, Brown University, Providence, RI, USA

Which factors were most important in determining the outcomes of major environmental justice cases in Louisiana during the 1990s? In four cases studied here, local citizens mobilised and engaged in protracted struggles. In the successful cases, protestors took advantage of a combination of circumstances and strategies that served to raise their group's power to levels that could challenge the opposing growth coalition, pursuing a political strategy targeting federal agencies that were in the process of creating policy, 'marketing' the struggle to attract outside resources and support, using the media and outside allies to put pressure on federal agencies, and seeking non-displacing goals. Fighting the siting of new facilities is far more likely to succeed than fighting existing contamination and entrenched firms.

Introduction

For decades now epic battles have been waged between environmental and community activists on the one hand and industry representatives, government officials, and other development advocates on the other. In some of these cases, activists have successfully stopped the siting of nuclear waste facilities or the dumping of toxic waste. In other struggles, industry representatives have won the right to build facilities or maintain landfills. The varied outcomes of these struggles raise the question of what factors are key in contributing to the success or failure of local activist groups. We re-assess the causes and impacts of four fairly prominent environmental justice struggles that took place in Louisiana. Two cases where citizens and environmental justice advocates fought the siting of new hazardous facilities – the Louisiana Energy Services

(LES) uranium enrichment plant siting in Claiborne Parish, and the Shintech PVC (polyvinyl chloride) factory siting in Convent – ended up as successes self-declared by the activists. In the two other cases, where activists sought the clean-up of existing contamination – the Grand Bois oilfield waste facility contamination case and the Agriculture Street landfill Superfund site in New Orleans – citizens gained very partial concessions and expressed ongoing frustration at the outcome.

To explore these cases, we provide very brief overviews of each and then look for the common characteristics that contributed to a protest group's success or failure. We realise that a comparison of siting and contamination cases is like comparing apples and oranges. However, we propose that this comparison is useful in an effort to illuminate the key differences between each type of fruit (or case). Other researchers have found that siting cases are more easily won by protestors than contamination cases, but we want to know why.

All of the struggles that we explore involve grassroots, poor and people-of-colour groups who are fighting against environmental injustice. This is not surprising since the cultural, political, and economic history of Louisiana has created a situation in which the populations most affected by the negative effects of development are poor people of colour (see Roberts and Toffolon-Weiss 2001). The proposed uranium (LES) and PVC (Shintech) plants would have had the greatest impact on the poor rural black communities closest to those two facilities. The massive oilfield waste pits in Grand Bois most endanger the poor people of Houma Indian and Cajun descent who live right next door. The Agricultural Street Landfill's potential risk is to the low and middle-income blacks whose homes were built directly atop it.

It is important to remember that these cases are only four among many battles over environmental injustice waged in the United States over these past three decades. Their outcomes were influenced by all those battles that came before them, when both environmental justice activists and business organisations learned many lessons, and during which laws and enforcement practices shifted (Bullard 1999). Political and cultural contexts shift: for example, the environmental justice movement received a significant boost in 1994 with President Clinton's Executive Order 12898, which decreed that 'all communities and individuals, regardless of economic status or race are entitled to a safe and healthy environment' (Environmental Protection Agency 1999). The order's impact remains ambiguous after eight years of neglect by the Bush administration since 2000, but it was instrumental in providing a legal and moral basis with which to pursue state and federal action to protect poor and minority communities through the latter half of the 1990s. In the 1990s, community groups throughout the country cited this order and filed complaints using Title VI of the 1964 Civil Rights Act with the Environmental Protection Agency (EPA) claiming that they, as minorities, were being disproportionately affected by pollution.

Assessing why some protests succeed

The first issue when studying the factors that contribute to a successful protest challenge is how to define 'success'. Social movement scholars have extensively explored how movement outcomes are classified. However, for present purposes we employ a simplified definition of success. A protest will be deemed successful if the protestors achieved the land use that they desired at the outset of the protest. In the Shintech and LES cases, the success of the protestors is quite obvious: neither unwanted land use was built in the community. For the other two cases, the protestors' initial goals – to be relocated in one case and to have the facility permanently shut down and cleaned up, in the other – were never realised. The contamination case outcomes are slightly more complex and difficult to classify because their end may not be fully realised, but in both cases, after many years of protest there are still some minimal ongoing efforts by the protestors to seek a resolution.

Researchers have identified several variables that affect social movement outcomes, such as organisational structure, resource levels, protest tactics, framing of grievances, political opportunities, and cultural opportunities. Some of these variables are under the control of actors, while others describe the structural environment in which the struggle takes place. We will briefly review these key concepts.

The structure of a protest group's *organisation* is one factor that can affect the social movement outcome. Tilly (1978) and McAdam (1982) emphasise the importance of a protest group taking advantage of an existing organisational structure. Several social movement researchers (Steedly and Fioley 1979, Mirowsky and Ross 1981, Gamson 1990, Frey *et al.* 1992), with the exception of Piven and Cloward (1977), found that groups with a unified, centralised, and bureaucratised organisation were more successful than those groups that had more decentralised organisational structures.

A protest group must make many strategic decisions regarding their *tactics*. Many researchers have found that more successful protest organisations use tactics that are disruptive or threaten disruption (Astin *et al.* 1975, Tilley *et al.* 1975, Steedly and Foley 1979, Mirowky and Ross 1981, McAdam 1983, Gamson 1990, Tarrow 1994). However, several studies of strikes and labour conflicts have found that the use of violence by protestors did not help them meet their goals (Taft and Ross 1969, Snyder and Kelly 1977). After an extensive review of contentious political battles, Giugni (1998) concludes that there is no definitive answer to the benefit or detriment of using disruptive tactics in a struggle.

Several studies have found that the use of *a legal strategy* combined with other strategies, such as demonstrations and lobbying, by protestors (including animal rights, pay equity, Indian rights, and civil rights groups) can lead to a successful outcome (McAdam 1982, Morris 1984, McCann 1994, 1998, Silverstein 1996). McCann theorises that the use of a legal strategy by a movement organisation can compel opponents to make concessions because of the fear of high legal fees and, ultimately, losing control of decision-making concerning the issue at hand. He also argues that legal tactics have not been successful for all groups.

Two options for legal strategy are a personal injury lawsuit or a public interest lawsuit. Environmental justice personal injury cases are often very difficult to prove. The plaintiff's case is frequently hindered by a lack of baseline health data in order to compare the health of the victims before and after exposure to a pollutant. Poor people usually have limited access to medical care, and, thus, lack good medical records. Government and corporate officials often do not take the symptoms of these people seriously, attributing the poor health of these communities to unhealthy lifestyles (e.g., eating fatty foods, smoking, drinking alcohol, and doing drugs) (Foreman 1998). Undertaking comprehensive health assessments is expensive and lengthy and there is often not enough money to pay for them in state and federal health departments. Further, middle-class jurors' own prejudices may affect their judgements when viewing poorer individuals with different life experiences (Kanner 1999).

Finally, a major hurdle that the environmental justice plaintiff faces is the lack of scientific evidence on the causal connection between certain exposures and health problems. It has taken years for scientists to draw a causal connection between lung cancer and smoking. These environmental cases involve toxins that have received much less scientific attention and there is often less specific information on the level and frequency of exposure. Additionally, there may be multiple sources of industrial pollution that have affected the plaintiffs and little may be known about their interactive effects. Technical assistance with medical and scientific evidence can be costly and, even if the plaintiff can afford it, it is likely that corporate defendants will be able to produce their own scientific experts to refute the plaintiff's argument (Kanner 1999). The other legal option is securing the services of a public interest lawyer and filing an environmental justice complaint or suit. The success of this type of strategy is highly dependent on the political and judicial climate at the time.

A social movement organisation must select specific *goals*. Research by Gamson (1990) suggests that single focus groups tend to be more successful. McAdam (1982) argues that disruptive goals that overtly challenge the existing political and economic structures of society will evoke a strong response and possible repression, while reform goals that seek incremental change may be less threatening, but also less productive. Gamson (1990), Frey *et al.* (1992) and Steedly and Foley (1979) argue that groups whose goals required the 'displacement' of opponents tend to be less successful that those with 'non-displacing' goals. The four cases we studied strongly support this point.

Another important decision that protestors must make is how to '*frame*' their grievances. The term 'collective action frame' is used to describe the collective beliefs and meanings that are developed by protestors and used to motivate and legitimate their protest (Benford and Snow 2000). Strategic framing of the problem/grievance, the solution, and 'a call to arms', can serve to attract additional protestors and bring in outside resources and support (Benford and Snow 2000). Both Walsh *et al.* (1993) and Gordon and Jasper (1996) found that the strategic framing of protest ideology to appeal to a wide audience may have significantly influenced the positive outcome of grass-roots protest against an incinerator siting. Similarly, Capek's (1993) analysis of a Texas superfund site provides evidence that the adoption of an effective environmental justice frame led to the success of grass-roots mobilising for a federal buyout and relocation.

Early social movement researchers have stressed that a protest group must have adequate *resources* to mount a successful challenge (McCarthy and Zald 1976, Cress and Snow 2000). While the availability of resources is not entirely under the control of protestors, they can develop strategies and tactics to obtain resources from outside supporters. Albrecht *et al.* (1996) noted that extra-local groups were instrumental in supporting local groups that were protesting against radioactive waste facilities by providing strategic, financial, moral, and informational assistance. Kitschelt (1985) also noted the importance of resources for the success of anti-nuclear movements in four different countries.

Structural factors such as political and cultural opportunities directly and indirectly affect the actions, motivation, and aspirations of the actors on both sides in the struggle. The growth coalition and the insurgent groups act within these structural constraints to develop mobilisation structures, frames, and strategies to increase their capital, and thus their power to affect economic development within their community. Amenta *et al.* (1999) recognises this link when he argues that the strategies of activists must fit into the current political context in order to be successful.

Political opportunities can be created by broad social processes that undermine and create instability in the political structure that serve to elevate the position of a minority group. These occurrences provide an opportunity for insurgent groups to gain political leverage and standing and decrease their susceptibility to repression (McAdam 1982, McCarthy 1996). Political opportunities are also created or diminished by legal precedents rendered by the outcomes of previous movements (McCann 1998).

We did not attempt to conduct a rigid comparative analysis and develop a model for social movement success. The cases do not lend themselves to this task because siting cases are fundamentally different from contamination cases. Instead we will analyse the siting cases together and then the contamination cases to determine why, as found in other studies, siting cases appear to be easier for protestors to win as compared to contamination cases.

Two new sitings and two existing contamination struggles

From 1995 through 2001, we conducted field and archival research on these four cases and the broader context of environmental justice struggles in

Louisiana.[1] We interviewed activists, industry representatives, and government officials on each case, and attended numerous meetings, protest events, and hearings as participant observers. We have assembled over 700 newspaper and magazine accounts of these local struggles, the industries they battled, and the environmental justice movement in general, together with hundreds of social movement organisational pamphlets and reports, company materials, and government documents.

Case 1: The nation's first major environmental justice judgment: the LES uranium enrichment facility

On 9 June 1989, US Senator J. Bennett Johnston announced in northern Louisiana's Claiborne parish that the area would be the site of a $750 million 'uranium enrichment facility', which would be entirely safe, and would bring to the area 400 jobs and millions in tax revenues. He proposed that the construction jobs would be plentiful and the tax revenues enormous for the depressed rural community. The cheerfully-named 'Claiborne Enrichment Facility' would be run by a group of investors called Louisiana Energy Services. This was a German-led consortium that included British and Dutch interests, as well as utilities from North Carolina, Minnesota, and Louisiana.

A series of informal gatherings had already taken place with wealthy residents and owners of local businesses. The town council, industrial development group, the police jury (roughly equivalent to county commissioners elsewhere), the state representative and state senator for the district, the congressman and senator were all on board for this project. The people that Johnston, the political establishment and the developers failed to even consider were those who lived closest to the proposed site. Five miles outside of the county seat, Homer, are Forest Grove and Center Springs, two small 100-year-old rural African-American communities connected by a narrow road upon which the enormous factory was designed to sit. The residents of those communities were neither at the announcement nor were they invited to the gatherings in the weeks beforehand.

These residents formed a protest group to fight the proposed land-use. They viewed the promised tax revenues as too good to be true, and believed that their poor communities were being 'sold out' for the benefit of local businesspeople and politicians. This tiny group, Citizens Against Nuclear Trash, or CANT – 'as in you CANT build it here' – eventually secured the legal support of EarthJustice, Greenpeace, the Nuclear Information Service (an established Washington-based anti-nuclear group), and several other local, national and international groups. Their coalition crossed race and class lines, and survived nearly a decade of battle before the matter was finally settled.

The protestors took advantage of several state-wide and national political opportunities. At that time in Louisiana, the head of the state Environmental Protection Agency, Paul Templet, was sensitive to environmental issues, and there were divisions within the federal government over whether the siting represented an environmental justice case. The EPA's regional administrator Jane Saginaw castigated the Nuclear Regulatory Commission (NRC) impact statement for not considering environmental justice (News-Star 1994). The agency was under pressure from the Clinton administration to consider the President's Executive Order on Environmental Justice (and many speculated that the agency was interested in keeping out private competition to its uranium monopoly).

The protestors used several strategies in their fight against the corporate giant. Members testified in front of Congress in hearings related to the struggle. The group's legal representation brought claims involving racist intent and impact of the proposed siting to the NRC, the state environmental agency, and the EPA. The group also conducted extensive public relations work. Many letters were written to the local newspaper and state and federal agencies explaining the protestors' views. They even sent letters to many lenders on Wall Street to warn them of the project's shaky financial foundations (Minden Press-Herald 1992).

The battle over LES swung back and forth in the following years. Finally on 2 May 1997, the NRC released a 'partial' ruling on the plant which was 'widely viewed as a national precedent in the area of environmental justice' (Shinkle 1998). In the ruling, the NRC cited evidence that 'racial considerations had played a part in the site selection process'. They also stated explicitly that they were addressing the case in the spirit of Clinton's Executive Order on Environmental Justice. LES finally put out a terse but angry press release that 'LES officials today ended their seven-year quest for a license from the U.S. NRC to build and operate what would have been the nation's first privately owned centrifuge enrichment plant.' The environmental justice coalition in this case had been successful, it seemed, partly by utilising what was evolving into a strategic delaying tactic, which made the project economically unfeasible by tying up its permits in the courts.

Case 2: EPA's environmental justice test case: the Shintech PVC plant

On 25 October 1996, the New Orleans *Times Picayune* newspaper published an article announcing the plans of the American subsidiary of a Japanese manufacturer to build a $700 million chemical plant on the Mississippi River in St. James Parish, a rural community located halfway between Baton Rouge and New Orleans on the banks of the Mississippi River. The article announced that the project would bring 2000 temporary construction jobs and 165 permanent jobs to the small town of Convent, Louisiana. The plant, which would be built on 500 acres of a former sugar cane plantation, would manufacture several chemicals, including PVC. The article reported that the St. James Parish officials had been pursuing the project since 1995 and were pleased by the company's decision to locate in their parish (Judice *et al.* 1996).

Strong and resolute groups of individuals with views on either side of the issue quickly formed. The pro-Shintech alliance was composed of company representatives, state and local business groups, the state Department of Environmental Quality (DEQ), the state Department of Economic Development, state and local elected officials, the local chapter of the National Association for the Advancement of Colored People (NAACP), and a group of local community residents. Many of these actors had been working with Shintech for months before the public announcement of the siting to ensure a smooth and quick siting process.

The alliance of groups that opposed the plant began with a single citizen's group formed by local community residents in August 1996. The group, St. James Citizens for Jobs and the Environment (SJCJE), was composed of approximately 100 official members, three-quarters of whom were African Americans. Two and a half years after the initial group formed, the anti-Shintech alliance grew to include 22 groups including state and national environmental and civil rights groups. The slogan of the opposition alliance – 'Enough is enough' – refers to their perceptions that there is more than enough pollution already in the area and that residents will not put up with any more contamination. The Tulane University Environmental Law Clinic aided the opposition coalition immeasurably by providing free legal expertise.

The group took advantage of numerous political opportunities in the form of inconsistent views on the case proposed by a development-focused state governor and the EPA, which was focused on implementing President Clinton's 1994 Executive Order on Environmental Justice. The protestors appealed to the EPA to rein in the state Department of Environmental Quality and make them comply with anti-pollution and environmental justice regulations. The coalition of activists also put pressure on the EPA by attracting national media attention and asking the general public to petition the agency on their behalf.

Initially, the state DEQ issued two of the necessary environmental permits for the plant's operations. However, the protestors and attorneys from the Tulane Environmental Law Clinic filed numerous complaints with both the EPA and the courts regarding violations in the environmental permits, inappropriate actions of parish and DEQ officials, and a claim of environmental injustice. The environmental justice complaint asserted that by granting air and water permits for the plant, the state DEQ was allowing a poor black community to be exposed to disproportionately high levels of pollution (Giordano 1997a, b). Responding to the complaint of an alleged violation, the EPA revoked Shintech's air permit. Additionally, the agency began an investigation into the environmental justice complaint.

The plant siting was tied up in legal proceedings filed by the Tulane Environmental Law Clinic for two years. Not knowing how long the legal battles would prevent the company from moving ahead, Shintech decided to greatly reduce the size of the plant and move its proposed location to Plaquemine, Louisiana, approximately 25 miles north of the original site, to a site adjacent to one operated by its close partner, Dow. Anti-Shintech forces viewed the company's decision as a decisive victory.

Case 3: Media savvy Cajuns and Houma Indians: fighting an oilfield waste dump in Grand Bois

Over a period of a week and a half in 1994, 81 white tanker trucks moved oilfield waste from an Exxon oil dumping site at Big Escambia Creek in the state of Mississippi to the massive waste dump pits of Campbell Wells, where millions of barrels of waste had been dumped over a span of nearly two decades. The smell from Exxon's waste wafted across Highway 24, where a passing woman trucker was sickened by it. The fumes then drifted through the tiny town of Grand Bois, home to 300 residents. Children coming home from school dashed from their school bus to their homes, covering their faces with their shirts or a handkerchief. Startled, other residents of Grand Bois repeatedly called the state's Department of Environmental Quality. Paul Muhler, staffing the department's emergency hotline, was sent down from the DEQ to Grand Bois.

Finding no help from the state Environmental Agency, the outraged residents appealed to a doctor from the region who had recently been elected to the state Senate to change the laws in the state, to regulate oilfield wastes and, more immediately, to close the pits at Campbell Wells. The residents also sought the assistance of a young trial lawyer from New Orleans, Gladstone Jones. The group received national media attention when CBS 60 Minutes reporter Ed Bradley and his 'CBS Investigative Reports' news team followed the story of Grand Bois for over a year, finally showing the hour-long special to a national audience just before Christmas 1997. Assistance from outside groups and politicians, however, was limited to the local state senator, the trial lawyer, the 60 Minutes report, the state-wide environmental action network, and a dedicated local scientist.

The efforts of the lone state legislator met with little success. He introduced a bill in the state Senate to have the US Liquids site in Grand Bois closed. The industry argued that the bill would shut down the oil drilling in the state by crippling its ability to access disposal sites. The 60 Minutes exposé put pressure on the Governor to investigate the allegations of contamination, but the ultimate decision-making for the case occurred in a state courtroom. Starting at the beginning of May, the Commissioner of Conservation at the Department of Natural Resources (DNR) issued a series of emergency rules because of 'public concern'. The new rules called for the testing of wastes at the site of production. However, oil industry advocates pressed for a reduction in the number of *types* of waste that required testing. Ultimately, only simpler tests were required, which one critic called 'very fudgeable' (Louisiana Department of Natural Resources 1999).[2]

Besides testing the waste, Governor Foster also sought more human health testing to confirm his suspicions that nothing was wrong in Grand Bois. Foster and the US Congressman from the district, Billy Tauzin, asked the US Centers for Disease Control and Prevention (CDC) to provide immediate assistance to the state agencies conducting the health assessment (Zganjar 1998). Grand Bois residents, however, decided they would be tested only by the scientists and agencies they trusted, and that specifically excluded the state Office of Public Health.

Less than a month after the Exxon trucks dumped their waste at the US Liquids ponds in Grand Bois, the community had filed a class action lawsuit. The massive case would take them three years and hundreds of thousands of dollars to prepare. During the trial, US Liquids and Campbell Wells decided to make the residents an offer to settle, which, though 'sealed', was reported to include damages payments of some $7 million, but this was spread over all 301 plaintiffs. The lawyers received 40% and expenses, a common commission for taking this kind of high-risk case. The average to the Grand Bois residents, therefore, was less than $14,000 per person. Finally, US Liquids agreed to make some important changes in how the facility would be operated in the future, the most important point for some of the plaintiffs. Exxon, however, waited for a verdict. The jurors decided that Exxon was not negligent in its handling and disposal of waste and that it needed only to pay approximately $30,000 in reparations to three residents. Exxon declared victory. Asked if they would appeal the tiny penalties, Exxon's attorney said no. 'We feel Exxon is vindicated absolutely' (McMillan and Dunne 1998). The anger of the community of Grand Bois continues to simmer since they did not get what they wanted: the closure of the contaminating site.

Case 4: The politics of living on a superfund site – Agriculture Street landfill

From 1910 to 1960 the City of New Orleans operated the Agriculture Street landfill site for operation as the city's municipal waste dump; waste was frequently incinerated on site and buried in the surrounding area. Nearby residents complained throughout the 1940s and 1950s of a terrible stench wafting from the landfill. By the time it was closed, the landfill was 17 feet deep and covered 95 acres downriver from New Orleans' famous French Quarter.

A decade later the Housing Authority of New Orleans (HANO) and the federal Department of Housing and Urban Development (HUD) first chose Agriculture Street as the site of the Press Park neighbourhood, consisting of 167 public housing units. The neighbourhood expanded in 1975 when a newly formed group called the Desire Community Housing Corporation (DCHC) submitted plans to construct 67 single-family homes and an elderly care facility. The DCHC completed the construction of these properties, which they named Gordon Plaza, and the Gordon Plaza Elderly Housing Apartments, using $7 million in federal funds from HUD in 1981 (Daugherty 1998).

Once in, residents quickly noticed that their new homes were a far cry from the American Dream they were promised. Shoddy construction of the houses and landfill debris in the yards foreshadowed impending problems related to living atop a landfill. One resident found the corpse of a cow in her front yard. Another resident found a rusted car door in her garden. In addition, almost all of the ground surrounding the residents' homes contained broken glass and other debris (Daugherty 1998).

After discovering that the land beneath and surrounding their homes contained dangerous contaminants, some of the most worried residents met with city officials in 1985 to discuss possible relocation of their community. In their struggle for relocation, the residents of Agriculture Street went on to enlist the support of numerous politicians, including New Orleans Mayor Marc Morial, Congressman William Jefferson, and United States Senators John Breaux and Mary Landrieu. The EPA came to inspect the site in May, 1986, taking soil samples, some of which had lead concentrations greater than 1000 parts per million (ppm), and three samples had lead concentrations of more than 4000 ppm, all far above EPA 'safe levels'. Soil samples also contained lead, zinc, arsenic, mercury and cadmium. There were polynuclear hydro-carbons, potentially dangerous oil products, in almost every sample (EPA 1999).[3]

The EPA, however, initially determined that the site was not dangerous enough to secure Superfund status and the federal monies to relocate the residents and clean up the site that might come with such designation. The neighbours and their advocates continued to fight, with the assistance of a regional tenants' rights organisation, the Gulf Coast Tenants Organization. The organisation put pressure on EPA to reconsider their methods of ranking sites. Congressman William Jefferson brought his office's weight to bear with the help of the Congressional Black Caucus. Then, in 1990, EPA rules changed to include soil contamination in hazard ranking scores, changing the neighbourhood's score to 50 out of 100, much greater than the 28.5 needed for Superfund designation (Cooper and Warner 1995).

The Agriculture Street residents were caught in the trap that has ensnared dozens of Superfund contaminated sites around the country: their neighbour-hood was certified as hazardous enough to get listed, but not hazardous enough to get them relocated. Their homes were worthless. The Superfund programme, meanwhile, was embroiled in a political battle between Democrats and the Reagan/Bush administrations. Many environmentalists feared that the Republican administrations were trying to subvert the programme (Szasz 1993).

The residents' efforts and frustrations were numerous. Faced with unsatisfying responses from government officials after a decade of pleas, some residents believed that only lawsuits could relocate them from atop the dump at Agriculture Street. For years the community had difficulty getting good public interest lawyers to take up their case, and the private lawyers they found have been interested only in class-action lawsuits, which have shown virtually no progress.

But the EPA did not believe that the Agriculture Street site was dangerous enough to warrant relocation, and instead opted to remove two feet of contaminated soil from residents' yards, replacing it with a 'geotextile' (a fine porous plastic mesh) barrier and two feet of 'clean' soil, at a cost of approximately $20 million. With the two-foot barrier created by the EPA cleanup plan, all trees in the neighbourhood were removed and residents were instructed on which trees they can plant. The plans also restrict residents from making additions to their homes and from building in-ground swimming pools. The residents tried several legal strategies trying to prevent the EPA from doing what they perceived as an inadequate clean-up. Legislators all unsuccessfully attempted to get the budget appropriations needed for the buy-out of the community.

Despite residents' protests and opposition from the City of New Orleans, the EPA began clean-up of the site in late 1998, piece by piece. Overall, protestors view their struggle as a failure because they have not received what they have wanted all along: relocation.

What was important for success?

The two siting cases are remarkably similar in many respects. In both cases, protest erupted against a proposed siting that had been planned by the companies long in advance; however, the residents who would be most immediately affected, who became the protestors, were never consulted. Both protest groups were well organised with leaders who remained focused on a single, non-displacing goal – to prevent the sitings. Having non-displacing goals is one of the key differences between siting cases and contamination cases (see also Gladwin 1987). Courts and administrators may be more reluctant to shut down an already existing facility and displace development and guaranteed tax revenues than to address a proposed siting. And, in fact, the courts or federal agencies did not have to actually reject the proposed sitings; in both of the successful cases, the companies, frustrated with the long legal delays, decided to withdraw their own proposals. The EPA actually heralded the Shintech resolution as a non-regulatory negotiated solution.

The fact that both groups framed their struggles as environmental justice battles served them well. Because this master frame was popular and came into the public spotlight with President Clinton's 1994 Executive Order, the two protest groups were able to attract support from state-wide and national environment and social justice groups and media attention. The LES and Shintech protestors accepted this assistance and applied the resources in an effort to put pressure on federal agencies to stop proposed facilities that were permitted by the state authorities. Both groups sought out public interest lawyers who would take their cases pro bono, rather than private 'personal injury' lawyers who charge large percentages of the final settlement amount *if* they win their cases in courts of law. With the assistance of public interest lawyers, the winning groups by contrast filed environmental justice complaints with the federal Environmental Protection Agency and various other lawsuits related to the cases.

The context of these cases is important: these struggles all occurred in Louisiana in the late 1980s and 1990s. The LES case was successfully resolved in early 1998, the Shintech case ended later in 1998, followed by the Grand Bois court proceedings in the same year. The Agriculture Street landfill case mostly was fought during this same period, but the story goes on as the site was badly flooded during hurricane Katrina in 2005, leading to widespread fears of toxic contamination. The close geographical and temporal proximity of these cases has some significance in terms of 'cultural opportunities'. First, the Shintech protestors saw and learned from the success of the LES activists. In fact, after their victory the LES protestors came down to Convent, Louisiana and met and celebrated with the Shintech protestors. These visits, organised by the Louisiana Environmental Action Network (LEAN) and EarthJustice, buoyed the spirits and informed the strategies of the Shintech protestors. Grand Bois and Agriculture Street protestors also attended LEAN events and gained knowledge from the others.

The timing of the cases is very important for another reason – political opportunity. The successful siting cases occurred at a time when the federal government was struggling to create regulations and policy for dealing with environmental justice cases. At this time, these agencies were more vulnerable to pressure from the public, media, and lobbying groups. It was the indecision and delay in developing and interpreting policy on the part of the Nuclear Regulatory Agency and the Environmental Protection Agency that ultimately caused the private companies to withdraw their plans. The hardening of business resistance to these environmental justice claims, and the election of George W. Bush in 2000, made for a much tougher context in the next decade.

In the two successful siting cases, the groups received a large amount of financial and organisational support from like-minded outside organisations. The outside support helped to focus public opinion and sway political leaders to support their cause and put pressure on the agencies that were handling the two cases. These activist groups all had elite allies, an open political system, and instability in the political system at different levels of government. Both the LES and Shintech cases occurred around the time of President's Clinton's Executive Order on Environmental Justice and were decided during the Clinton Administration at a time of increased openness to environmental protection. This prompted the EPA and other federal agencies to take environmental injustice complaints seriously, thus creating a division between the federal government and state-level political actors, a division that the protestors exploited.

While the public interest lawyers were pursuing their environmental justice complaints, it was equally important to have the protestors getting out the message regarding their struggle in an effort to have the media, lobbying groups, and the general public put pressure on the federal agencies. When agencies know they are being watched they may be more careful regarding their decisions. These political openings that advantaged the protestors in the LES and Shintech cases closed soon after the Shintech case was resolved. Since then, the EPA and the Louisiana Supreme Court have demonstrated a more conservative approach to environmental justice and several court decisions have limited the usefulness of civil rights legislation for environmental justice claims.

The legal strategies chosen by the protest groups were consistent with failure or success. The protest groups in both the Shintech and LES cases tried to pressure the federal agencies to develop environmental justice policy and regulations related to President Clinton's Executive Order and Title VI. Both communities secured public interest lawyers, rather than relying on private personal injury lawyers and a class action 'toxic tort'. Although the government did not directly decide in the residents' favour, this strategy resulted in substantial delays for the siting of the two plants that caused the companies to decide to build elsewhere (see Walsh *et al.* 1997). The successful struggles were fought in the national political arena as opposed to the courtroom (see also Cole and Foster 2001 on this point). This suggests a limitation in relying on legal strategies to resolve environmental justice cases. Especially troubling are the poor outcomes of class action lawsuits: even when they win, communities and families can be torn apart by jealousy and distrust. However, more frequently their cases are deferred for years by distracted lawyers and by the legal proceedings forced by corporate or government lawyers.

The trajectories of the failed contamination protests were very different from the ones outlined above, although there are factors in each case that are similar to the victorious siting cases. These protestors were facing situations of existing contamination. In one case, unsuspecting residents experienced a sudden incursion of toxic fumes; in the other, residents gradually became aware that they were living on top of a landfill that was seeping to the surface. In both of these instances the residents first sought the help of local and federal officials. The Grand Bois residents solicited assistance from the state agency for environmental quality and, then, when that failed, they requested help from their Congressman. The Agriculture Street group sought relief from the EPA Superfund Program. Neither of these approaches gave the protestors what they wanted.

The resources available to the Grand Bois protestors were limited partly due to the fact that they did not frame their struggle as an environmental justice issue. The limited support they received from the 60 Minutes television exposé on their situation dissipated not long after it was aired. However, similar to those in the siting cases, the Agriculture Street protestors did frame their struggle as an environmental justice issue and they received outside support from state-wide and national organisations. But, this was not enough. Although three other Louisiana communities had been relocated, these relocations were financed by the chemical companies that were located next door. The Agriculture Street residents' misfortune in not being able to pressure an institution with 'deep pockets' was that it was not a rich corporation but New Orleans – one of the poorest municipalities in the USA – that was responsible for the contamination.

Finally, the protestors in the failed cases resorted to hiring private lawyers to file lawsuits against the owners of the source of the contamination. As stated earlier, this approach is often fraught with problems for the plaintiffs. In both cases, the protestors received little relief from this legal avenue, in spite of the remarkable efforts of the Grand Bois community's lawyer.

Since the time of these four cases, the political climate affecting environmental justice struggles continues to change. It changed from one of open political opportunities for environmental justice claims to a more closed arena. After the resolution of the Shintech case, the EPA worked more quickly and decided one of the first substantive decisions regarding a Title VI complaint against Michigan Department of Environmental Quality's Permit for the proposed Select Steel Facility; in this case the EPA determined that there was no adverse impact on the minority community consistent with Title VI and it dismissed the complaint (http://www.epa.gov/civilrights/recdecsn.htm [Accessed 17 July, 2008]). This was a major step backwards from the Shintech case, where the EPA stated in a preliminary report that 'disproportionate impact' should be considered regardless of whether the permitted discharges were within the letter of the law for single facilities. Select Steel decided to build in Ohio instead, but the spectre of 'Shintech-style' permitting analyses was apparently beaten back.

By the end of the Clinton Administration, the EPA had not finalised the guidance for federal agencies on environmental justice cases. The EPA under President George W. Bush leaned away from using civil rights laws to address environmental justice and towards using the National Environmental Policy Act (NEPA). This act, passed in 1969, has the purpose of establishing a national policy for the environment that promotes harmony between man and his environment. Former EPA Secretary Christine Todd Whitman pursued a less confrontational approach that may be more in step with the desires of many state environmental agencies, who fear driving away major employers (Yamada 2002). The EPA is still accepting Title VI administrative complaints filed by individuals, but the Supreme Court's 2001 *Alexander v. Sandoval* case has undermined 'disproportionate impact cases' and requires that plaintiffs prove intentional discrimination. Few cases are moving forward, and the EPA under Bush sought to promote 'alternative dispute resolution' out of the courts. Now, only a federal agency can enforce the Title VI, section 602 disparate impact regulations (Yamada 2002). Ultimately, the third circuit court decided that there was no congressional intent to create a private enforceable right under Title VI to be free from disparate impact discrimination. Now President Obama's choice to head the Environmental Protection Agency, Lisa Jackson, is an African-American woman from Louisiana. Although it is too early at the time of writing to conclude what impact the change in government will have on particular environmental justice cases, the political opportunity structure is far more positive than under Bush (see Mohai *et al.* 2009).

In summary, this analysis points to the complexities of environmental justice cases and the significant differences between siting and contamination cases. It is not just one or two characteristics of a protest that decide victory. Rather, in these cases, it was a combination of pursuing a political strategy targeting federal agencies that are in the process of creating policy, 'marketing' the struggle to attract outside resources and support, using the media and outside allies to put pressure on the federal agencies, and seeking non-displacing goals. The analysis of these four cases demonstrates that the siting cases were more likely to have the key ingredients that would lead to success as compared to the contamination cases. Although a protest group, such as in the Agriculture Street contamination case, had outside support and adopted an environmental justice frame, their legal strategy, which focused first on the EPA and then on a private toxic tort, was not enough to overcome a political environment that was closed to its demands. Similarly, the protestors in Grand Bois who wanted relief from open oilfield waste pits, found no success from their political efforts to change regulations and little relief from their toxic tort. They were pursing an uphill battle, with little outside support, in a political environment where oil is king. While the LES and Shintech cases have been widely cited as landmarks and pivotal cases nationally for the national environmental justice movement, recent judgments in the courts and redefinitions by agencies in Washington have put in doubt their lasting impact. A new administration brings some hope but little certainty about the direction of environmental justice claims in the USA.

Notes

1. For more detail on these cases and some of our arguments here, see Roberts and Toffolon-Weiss 2001.
2. The critic was chemist Wilma Subra, citing different positions the lid can be on the 'sniffer' machine. Interview, 1999.
3. Further testing by the Environmental Protection Agency in 1993 confirmed the 1986 tests indicating excessive levels of toxins such as lead, arsenic, chromium, and calcium. Tests also revealed the presence of 'volatile organic compounds, polyaromatic hydrocarbons, metals' and other dangerous toxins' (Environmental Protection Agency 1999).

References

Albrecht, S., Amey, R., and Sarit, A., 1996. The siting of radioactive waste facilities: what are the effects on communities. *Rural Sociology*, 61, 649–673.

Amenta, E., Halfmann, D., and Young, M.P., 1999. The strategies and contexts of social protest: political mediation and the impact of the Townsend Movement in California. *Mobilization: An International Journal*, 4 (1), 1–23.

Astin, A., *et al.*, 1975. *The power of protest*. San Francisco: Jossey-Bass.

Benford, R. and Snow, D., 2000. Framing processes and social movements: an overview and assessment. *Annual Review of Sociology*, 26, 611–639.

Bullard, R., 1999. *Environmental Justice in the 21st Century*. Atlanta, GA: Environmental Justice Resource Center. Available at http://www.ejrc.cau.edu/ejinthe21century.htm [Accessed 15 October 2009].

Capek, S., 1993. The environmental justice frame: a conceptual discussion and an application. *Social Problems*, 40, 5–24.

Cole, L.W. and Foster, S.R., 2001. *From the ground up: environmental racism and the rise of the environmental justice movement*. New York: NYU Press.

Cooper, C. and Warner, C., 1995. Is old dumpsite toxic? Not very, EPA says. *New Orleans Times Picayune,* 4 April.

Cress, D. and Snow, D., 2000. The outcomes of homeless mobilization: the influence of organization, disruption, political mediation, and framing. *American Journal of Sociology,* 105, 1063–1104.

Daugherty, C., 1998. Digging in. *The Gambit Weekly* (New Orleans), 3 November.

Environmental Protection Agency (EPA), 1999. 'Record of decision abstracts: Agriculture Street landfill. Available at http://www.epa.gov/envirofacts [Accessed 27 July 1999, EPA Record of Decision].

Foreman, C., 1998. *The promise and peril of environmental justice.* Washington, DC: Brookings Institution Press.

Frey, R.S., Dietz, T., and Kalof, L., 1992. Characteristics of successful American protest groups: another look at Gamson's strategy of social protest. *American Journal of Sociology,* 98, 368–387.

Gamson, W., [1975] 1990. *The strategy of social protest.* Homewood, IL: Dorsey Press.

Giordano, M., 1997a. Delicate chemistry. *New Orleans Times Picayune,* 27 May, p. B1.

Giordano, M., 1997b. DEQ clears way for PVC plant. *New Orleans Times Picayune,* 29 May, p. A1.

Giugni, M., 1998. Was it worth the effort? The outcomes and consequences of social movements. *Annual Review of Sociology,* 24, 371–393.

Gladwin, T.N., [1980] 1987. Patterns of environmental conflict over industrial facilities in the United States, 1970–78. *In:* R.W. Lake, ed. *Resolving locational conflict.* New Brunswick, NJ: Rutgers Center for Urban Policy Research, 14–44.

Gordon, C. and Jasper, J., 1996. Overcoming the NIMBY label: rhetorical and organizational links for protestors. *Research in Social Movements, Conflict, and Change,* 19, 159–181.

Judice, M., King, R., and Welsh, J., 1996. St. James to get $700 million plant; Japanese chemical company promises 165 permanent jobs. *New Orleans Times-Picayune,* 25 October, p. C1.

Kanner, A., 1999. Assisting injured individuals. *In:* M.B. Gerrard, ed. *The law of environmental justice, theories and procedures to address disproportionate risks.* Chicago: American Bar Association, 619.

Kitschelt, H., 1985. Political opportunity structures and political protest: anti-nuclear movement in four democracies. *British Journal of Political Science,* 16, 57–85.

Louisiana Department of Natural Resources, Office of Conservation, 1999. Declaration of emergency, amendment to Statewide Order No. 29-B (Emergency Rule), 29 May. Available at http://dnr.louisiana.gov/cons/conserv.ssi [Accessed 20 July 1999].

McAdam, D., 1982. *Political process and the development of black insurgency, 1930–1970.* University of Chicago Press.

McAdam, D., 1983. Tactical innovation and the pace of insurgency. *American Sociological Review,* 48, 735–754.

McCann, M., 1994. *Rights at work: pay equity reform and the politics of legal mobilization.* University of Chicago Press.

McCann, M., 1998. Social movements and the mobilization of law. *In:* A.N. Costain and A.S. McFarland, eds. *Social movements and American political institutions.* New York: Rowman & Littlefield Publishers, Inc, 201–215.

McCarthy, J., 1996. Constraints and opportunities in adopting, adapting and inventing. *In:* J. McCarthy, D. McAdam, and M.N. Zald, eds. *Comparative perspectives on social movements.* New York: Cambridge University Press, 141–151.

McCarthy, J. and Zald, M., 1976. Resource mobilization and social movements: a partial theory. *American Journal of Sociology,* 82, 1212–1241.

McMillan, J. and Dunne, M., 1998. Grand Bois plaintiffs shocked jury didn't see harm. *Baton Rouge Advocate Online.* Available at http://www.2theadvocate.com [Accessed 10 August 1998].

Minden Press-Herald, 1992. 'CANT coalition moves LES fight to new ground. *Minden Press-Herald,* 10 June.

Mirowsky, J. and Ross, C., 1981. Protest group success: the impact of group characteristics, social control, and context. *Sociological Focus,* 14, 177–192.

Mohai, P., Pellow, D., and Roberts, J.T., 2009. Environmental justice. *Annual Review of Environment and Resources,* 24, 16.1–16.26.

Morris, A., 1984. *The origins of the civil rights movement.* New York: Free Press.

News-Star, 1994. Agencies do battle over plant. *News-Star,* 5 February.

Piven, F.F. and Cloward, R.A., 1977. *Poor people's movements, why they succeed, how they fail.* New York: Vintage Books.

Roberts, T. and Toffolon-Weiss, M.M., 2001. *Chronicles from the environmental justice frontlines.* New York: Cambridge University Press.

Shinkle, P., 1998. Uranium plant plan dropped. *Baton Rouge Advocate,* 13 May.

Silverstein, H., 1996. *Unleashing rights: law, meaning, and the animal rights movement.* Ann Arbor, MN: University of Michigan Press.

Snyder, D. and Kelly, W., 1979. Strategies for investigation violence and social change: illustrations from analyses of racial disorders and implications for mobilization research. *In:* M. Zald and J. McCarthy, eds. *The dynamics of social movements.* Cambridge, MA: Winthrop, 212.

Steedly, H.R. and Foley, J.W., 1979. The success of protest groups: multivariate analysis. *Social Science Research,* 8, 1–15.

Szasz, A., 1993. *Ecopopulism: toxic waste and the movement for environmental justice.* Minneapolis: University of Minnesota Press.

Taft, D. and Ross, P., 1969. American labor violence, its causes, character, and outcome. *In:* H. Graham and T. Gurr, eds. *Violence in America.* New York: Praeger, 281–380.

Tarrow, S., 1994. *Power in movement: social movements, collective action, and mass politics in the modern state.* Cambridge: Cambridge University Press.

Tilly, C., 1978. *From mobilization to revolution.* New York: McGraw-Hill Publishing Company.

Tilly, C., Tilly, L., and Tilly, R., 1975. *The rebellious century, 1830–1930.* Cambridge, MA: Harvard University Press.

Walsh, E., Warland, R., and Smith, C., 1993. Backyards, NIMBYs, and incinerator sitings: implications for social movement theory. *Social Problems,* 40, 24–38.

Walsh, E., Warland, R., and Smith, C., 1997. *Don't burn it here.* University Park: The Pennsylvania State University Press.

Yamada, G.H., 2002. Redirection for environmental justice. *Federal Facilities Environmental Journal,* 13, 55–65.

Zganjar, L., 1998. Louisiana again seeks help in Grand Bois. *Baton Rouge Advocate,* 31 May.

More acted upon than acting? Campaigns against waste incinerators in England

Christopher Rootes

Centre for the Study of Social and Political Movements, School of Social Policy, Sociology and Social Research, University of Kent, Canterbury, UK

Campaigns against waste incineration in England never achieved prominence comparable with that of 1990s anti-roads protests. The explanation lies in the relative centrality of policies to government, the availability of allies, and the local nature of policy implementation and siting decisions. Variation in the outcomes of local campaigns is best explained by the differing political opportunity structures of local government. Historic patterns of local waste management, the timing of proposals and changes in government policy are also factors. Sharply rising costs of landfill drive waste authorities to seek alternatives, and new proposals for incinerators increased after 2005, provoking the establishment of a national anti-incinerator network. However, increased concern about climate change and availability of new, modular waste treatment technologies reduce the appeal of incineration.

Introduction

In the 1990s, Britain attracted interest internationally from environmental activists and commentators because of the vigour and innovativeness of protests against road-building. From Twyford Down in the south, to Glasgow Pollock in the north, Solsbury Hill in the west to the M11 motorway extension through east London, anti-roads protests raged, enjoyed widespread popular support and contributed to one of the most spectacular policy U-turns in modern British history as 'Roads for Prosperity', one of the cornerstones of the development policies of the Thatcher and Major governments, was abandoned.

With that example of an apparently successful anti-roads campaign so fresh in the memories of activists and the broader public, it might therefore have been expected that a new wave of infrastructure projects no less telling about the environmental costs of advanced capitalism and consumerism might also have excited vigorous, well-supported and well-networked protest campaigns. Waste, after all, goes to the heart of the issues of unnecessary and environmentally damaging production and consumption even more than does road-building. Yet although there have been many local campaigns against proposed waste incinerators and landfills, few have attracted national media attention even briefly, none has acquired an iconic status similar to that enjoyed in the struggle against road-building by Twyford Down or Newbury, and until very recently there has been no effective networking of the various local campaigns comparable with that developed in the course of anti-roads protests.

The story of campaigns against proposed waste facilities in Britain has until very recently been that of largely isolated campaigns in geographically dispersed parts of the country, occurring according to idiosyncratic timetables, attracting limited and/or intermittent interest and support from national environmental movement organisations and sustaining only desultory efforts at coordination from one campaign to another.

How are we to explain the striking failure of campaigns against waste infrastructure to develop in a way comparable to anti-roads protests? And how are we to account for the considerable variation in the outcomes of campaigns against proposed waste treatment facilities from place to place and over time? In answering those questions, we will consider the features of proposed waste infrastructure developments, the character of the sites for which they are proposed, the political opportunity structures and the changing policy contexts that frame them. First it is necessary to set the emergence of the waste issue in England in context.

Background and context

The UK government's draft waste strategy document, *A Way With Waste* (DETR 1999), laid out the background to the recent contention over waste management. In 1999, household waste in England amounted to some 25 million tonnes per annum, and was estimated to be increasing at an annual rate of 3%. Historically, most municipal solid waste in England has been consigned to landfill. In 1999, 85% of English household waste was sent to landfill, a proportion far higher than in other comparably industrialised countries; the proportion of domestic waste that was recycled (8%) or incinerated (7%) was correspondingly low.

Most landfill sites in England are voids created in past decades by quarrying or the extraction of clay for brick-making. As existing landfill sites filled up and as planning constraints and opposition from the public and conservation organisations such as the Council for the Protection of Rural England (now Campaign to Protect Rural England) (CPRE) made it increasingly difficult to find new sites, the unsustainability of England's management of domestic waste became increasingly apparent. Better understanding of the hazards associated with landfill stimulated tighter regulation and, to provide economic incentives to seek more sustainable alternatives, the UK government in 1996 imposed a landfill tax, which subsequent governments have progressively increased.

The urgency of finding alternatives to landfill was greatly increased by the EU Landfill Directive (Council Directive 99/31/EC), which requires the volume of biodegradable municipal waste sent to landfill to be reduced to 75% of the 1995 level by 2010, 50% by 2013 and 35% by 2020. Failure to meet these targets could result in fines of €270 million per year, and so government has handed down correspondingly demanding targets to local waste authorities. Although as early as 1990 government had set a 25% target for waste recycling, local waste planning authorities remained sceptical that recycling rates could be sufficiently increased, and even waste authorities that had never incinerated municipal waste began to search for sites suitable for incinerators. The published Waste Strategy 2000 (DETR 2000) did not privilege incineration over other means of waste disposal, but, as recycling rates remained stubbornly low, ministers appeared to accept that it might be necessary to build as many as 100 new incinerators in order to meet EU targets for the reduction of landfill.

Controversy over the construction of new waste incinerators had, however, already been stimulated by an earlier EU directive on incinerator emissions (1989 – EC Directive 89/429/EEC) which set limits on emissions from existing as well as new municipal waste incinerators. As a result, all the then-existing waste incinerators in England were closed or required extensive modification. In many cases, local waste authorities that had previously incinerated waste proposed to build new incinerators that complied with the Directive on the same sites as the old, polluting incinerators they were intended to replace. Surprisingly, whereas the operation of the old incinerators had generally been locally uncontentious, their replacement with new, much cleaner incinerators was sometimes fiercely resisted. Proposals for entirely new incinerators almost all encountered public opposition, and though the outcomes of the campaigns against them have varied, the number of new incinerators in England remains relatively small.[1]

There were, in August 2009, 18 operational large-scale waste incinerators in England. Three more were under construction. The website of the recently formed anti-incinerator network UK without incinerators network (UKWIN) (http://www.ukwin.org.uk) listed a further 60 waste facilities as proposed, subject to approval and funding. However, of these, many were at very early stages of consideration, and only a handful were cases in which the developers were committed to, and the planners had approved, conventional large-scale waste-to-energy incinerators; in many cases the technology remained to be determined, and in a number the plans were for gasification of waste in relatively small plants.

Yet there has been no prominent and sustained national campaign against waste incineration, local anti-incinerator campaigns have rarely been effectively linked one to another, and it was only in 2008 that attempts to establish a national network of local campaigns bore any very visible fruit. How then are we to explain why so few new waste incinerators have been built? Is it perhaps that, even in the absence of a sustained, highly visible national campaign, the cumulative impact of local campaigns has had the effect of pushing incineration down the list of preferred options for dealing with municipal waste? This is a question to which we shall return, but first let us consider why campaigns against proposed waste facilities never achieved the high profile of those against roads, and why the outcomes of local campaigns have varied.

Why campaigns against waste facilities never achieved the high profile of those against roads

Perhaps the principal reason that campaigns against waste facilities did not cohere and acquire as high a profile as did those against roads is that whereas road-building was a central policy of an increasingly unpopular government, national government's policy on waste has been relatively equivocal. Roads for Prosperity was an iconic policy of the Thatcher and Major governments, but there has never been a clear government policy favouring waste incineration. Moreover, while the implementation of roads policy was substantially national, the formation and implementation of waste policy has been distributed among the large number of local waste authorities working within the constraints of nationally fixed guidelines and targets but without detailed central prescription of the means by which those targets should be reached. Whereas the centrality of road-building to Conservative government policy made it a magnet for those who opposed the government and its vision of Britain's future, the lack of any comparably clear, high priority government policy for waste makes it an improbable rallying point for opposition.

The nature of the sites of the proposed developments also differed. New roads almost invariably traversed open landscapes hitherto spared the impact of industry, thereby increasing the industrial footprint and diminishing the already shrinking area of England remote from pollution by noise and artificial light, often compromising areas of special landscape value or sites of historic or special scientific interest. Waste incinerators, by contrast, were generally proposed for brownfield sites already blighted by industrialisation.

Moreover, whereas the similarities of roads projects and their close temporal coincidence aided networking of local campaigns, the variety of technologies and antagonists involved and the erratic timing of waste projects made it difficult to network them effectively. A major contribution to the networking of anti-roads protesters came in the form of the resources and networking capacities of Transport 2000, a lobby group originating in the railway unions, that, with the aim of promoting public transport, had established links with a variety of major environmental NGOs as well as local anti-roads groups (Dudley and Richardson 2001, pp. 158–159). Anti-incinerator campaigners, by contrast, have lacked powerful national allies and have enjoyed limited support from national environmental NGOs.

Greenpeace has campaigned intermittently against waste incineration, generally highlighting possible toxic emissions as part of its longstanding anti-toxics campaigning. It was drawn into national anti-incinerator campaigning by the prominence given to incineration in Waste Strategy 2000, and community groups quickly saw Greenpeace as a focal point, especially after Greenpeace staged a high profile direct action at the Edmonton, north London incinerator. To support its 'Incinerator Busters' campaign, Greenpeace appointed a staff member to support local campaigners and to coordinate Greenpeace activists in the regions. Greenpeace did not want to make a long-term commitment to coordinating a grassroots network, but preferred instead to support independent grassroots activists, many of whom were Greenpeace 'active supporters', and it was keen to get across its general anti-incineration message rather than to encourage NIMBY protests.[2]

The highpoint of Greenpeace's efforts to bring anti-incinerator campaigners together came in 2002 when, on Global Anti-Incineration Day, some 60 Greenpeace active supporters participated, as members of the Global Anti-Incineration Alliance (GAIA), alongside some 40 people from incinerator action groups from across the country in the occupation of a partly built waste incinerator at Basingstoke.[3] It is noteworthy, however, that this Greenpeace action was part of an international day of action, and that Greenpeace UK's other highly public anti-incinerator actions in 2001–2002 were directed at already operational waste incinerators in London and Sheffield rather than in support of local campaigns against proposed incinerators. Greenpeace appeared less interested in supporting embattled local communities than in implanting a network of local activists who would spread Greenpeace's own message.

When in 2002 the Strategy Unit report, which downplayed incineration as an option for waste management, was published, Greenpeace's campaign priorities shifted. Greenpeace withdrew from active campaigning on incineration, hoping that the local activists and groups with which it had made contact would sustain themselves as an autonomous network. But without the support of Greenpeace staff, individual activists were not prepared to take on the work needed to maintain a network. It appears that although Greenpeace had succeeded in networking anti-incinerator activists in various parts of the country and for a time supported them well with advice and campaign materials, it failed to create a network that could exist without continued Greenpeace support. When Greenpeace moved on, the embryonic network collapsed. Greenpeace had, in fact, not supported a network because one did not really exist; it had supported individual campaigners in the hope that they would attract others.[4] But although Greenpeace may not have been an organisation ideally structured to foster grassroots networking, it is doubtful that there was, at that time, a sufficient critical mass of local campaign groups capable of maintaining a self-sustaining network.

Although Friends of the Earth (FoE) has more consistently addressed the issue, it has rarely made it a high priority for national campaigning.

Nevertheless, from the mid-1990s, FoE supported local campaigners at public inquiries, held training days for anti-incinerator activists, and produced 'how to campaign' booklets designed specifically for anti-incinerator campaigners.[5] Its reputation for interest in the issue thus established, FoE has been an accessible point of contact for new campaigns, though it lacked the resources to give all the support the campaigners sought. FoE did not succeed in establishing a viable network of anti-incinerator campaign groups until, in 2008, it facilitated the formation of UKWIN, on whose board FoE has a representative.[6]

Until the appearance of UKWIN, networking among anti-incinerator campaigns was mostly achieved by the efforts of the more vigorous local campaigns and their increasing use of the internet to advertise their actions and to post links to other active campaigns, sometimes augmented by postings on environmental activist websites such as http://www.earthfirst.org.uk/actionreports/. By 2006, FoE was in contact with about 30 local groups opposing incinerators. With so many groups, FoE was finding it difficult to support and inform them individually. The decision to support formation of UKWIN was, therefore, partly a matter of resources and campaign priorities; with FoE's efforts increasingly focused upon climate change, there were not the resources to give continuing support to a large number of anti-incinerator campaign groups.[7] But its arms-length support of UKWIN also fit well with FoE's philosophy of empowering local communities.

One reason that national environmental NGOs have not more enthusiastically supported local campaigners against waste incineration is that their perceptions of the issues differ. Grassroots opposition to waste incineration generally focuses around concerns about air pollution and impacts upon health. While Greenpeace and local FoE activists have often been willing to embrace such concerns, the absence of clear scientific evidence to justify them has inhibited FoE's national officers from endorsing local anti-incinerator campaigns.[8] FoE's national anti-incineration campaigning has instead been the corollary of its promotion of recycling, labelling waste incineration a 'waste of resources', and latterly of its campaign against climate change.

A more general impediment to the networking of anti-incinerator campaigns is the land-use planning system. An incinerator proposal must pass through three stages: first, the local Waste Planning Authority must develop a Waste Local Plan (or Waste Development Framework) to establish the broad parameters of policy and criteria for the selection of waste disposal sites; second, the developer must submit an application for the construction of a particular facility on a particular site; third, an application must be made to the Environment Agency for certification that the project is compliant with local and national air quality and other environmental regulations. This third stage, taken by a national regulatory agency whose criteria are science-based, is effectively depoliticised with the result that almost all battles against waste incineration are local campaigns against particular proposals for particular sites according to locally idiosyncratic timetables. With the help of the internet, local campaigners may share advice and experience but, in the absence of a clear national policy in favour of waste incineration, case by case decision-making and the asynchronous timing of proposals for new incinerators severely limit opportunities for the development of a universal campaign. Only after 2005, when the number of incinerator proposals suddenly increased, did the critical mass of cases exist to sustain the formation of a national network of anti-incineration campaigners.

Nevertheless, if campaigns against waste incinerators have never developed into a movement comparable with that against road-building, it remains to be explained why some anti-incinerator campaigns have enjoyed success whereas others have failed.

Why some campaigns succeed while others fail

Various arguments have been advanced to explain why some campaigns against waste facilities have been more successful than others. These arguments have focused upon: the characteristics of the communities for whose localities the facilities are proposed; the skill, ingenuity, determination and resources of the campaigners; the mobilising capacity of the discourse employed by campaigners as opposed to the legitimatory capacity of the discourse employed by the advocates of such infrastructure projects; the political opportunity structures with which campaigners are confronted; the changing political and policy contexts within which decisions about waste management are made. We shall consider each in turn.

Community

Walsh *et al.* (1997) suggest that the outcomes of campaigns against waste infrastructure projects are more powerfully influenced by the skills, ingenuity, and determination of campaigners than the characteristics of communities. This conclusion perhaps reflects the fact that in the US, waste facilities were concentrated in or near poor neighbourhoods whose populations were disproportionately people of colour and/or were disadvantaged by limited command of English and/or lack of citizenship rights (Cole and Foster 2001). It is nevertheless surprising because the most general conclusion of political sociology is that political participation is patterned in accordance with the distribution of resources (see, e.g., Dalton 2008, p. 580). We should, therefore, expect that communities with high proportions of affluent, well-educated and otherwise politically resourceful inhabitants will more often sustain effective campaigns to defend their communities against the imposition of locally unwanted land uses than will places whose residents suffer from cumulative social and economic disadvantages and lack politically relevant resources.

This is generally the pattern with respect to anti-incinerator campaigns in England: socially diverse and resource-rich communities such as rural villages and small towns have generally been better placed to resist than those in urban areas with high concentrations of ethnic minorities and that concentrate social deprivation and political disadvantage. Thus neither the South East London Combined Heat and Power (SELCHP) waste incinerator project (Gandy 1994, p. 67), announced in 1989, nor the replacement waste incinerator for Birmingham Tyseley, proposed in 1993, was strongly resisted (Rootes 2006). The pattern is not uncomplicated, however.

Perhaps the most celebrated case of a successful campaign against a waste incinerator was that waged from 1999 to 2003 to prevent the construction at Byker in Newcastle-upon-Tyne of a new incinerator to replace an existing combined heat and power plant. That campaign was successful despite the fact that the Byker ward is among the 1% of most socially deprived wards in England (Dodds and Hopwood 2006). It is perhaps significant, however, that the Byker estate is bordered by newer private estates whose inhabitants are more affluent and more resourceful. Moreover, the Byker case is complicated by the longstanding complaints of residents about the smells and soot emitted by the old incinerator and by the revelation, at the very time that approval was sought for a new waste incinerator, that ash from the old incinerator, spread on local footpaths and allotments, had contributed to massive dioxin contamination of the soil. Thus political attention was focused upon the case and public confidence in the authorities was fatally undermined.

If the Byker case demonstrates that even deprived communities may successfully resist the imposition of incinerators, the experience of campaigns based on rural villages or small towns has also been mixed: whilst many have been successful, some have not. Clearly, the characteristics of local communities do not suffice to explain the pattern of successes and failures.

Discourse

Nor, however, does the ingenuity of local campaigners in employing universalising discourses. Walsh et al. (1993) argued that campaigns employing universalistic arguments were more likely to succeed than campaigns that employ particularistic, NIMBY arguments.

One compelling universalistic environmental discourse is that of ecological modernisation (EM). EM, recognising the structural character of modern environmental problems but seeking to address those problems without fundamental change to existing political, economic, and social institutions, has become 'the most credible way of "talking green" in spheres of environmental policy-making' (Hajer 1995, p. 30). EM proposes not 'end-of-pipe' solutions but redesign of industrial processes so that their impacts upon the environment may be minimised (Milanez and Bührs 2007) and, because it generally entails the minimisation of waste, it is usually held to offer 'win–win' solutions that save industry money as well as providing environmental benefits. As EM has become hegemonic, so the discourse of those who propose new waste facilities has changed. Mass burn incineration without energy recovery was now regarded as a mere 'technological fix' barely preferable to landfill, but incineration with energy recovery, suitably relabelled 'energy-from-waste' (EfW), was enthusiastically embraced as the ecologically modern alternative.[9]

For local communities opposed to the siting of waste facilities, however, universalist discourses such as ecological modernisation are less clearly advantageous, and there are a number of strategic dilemmas involved in attempts to raise the level of campaigns beyond the local and particular. Campaigns that adopt a universalising discourse might reframe local issues as tokens of broader questions of environmental management appealing to a larger public and so make it easier to recruit the support of non-local actors, but such universalistic arguments may appear overly abstract and complex to local people who may be more easily and intensely mobilised by NIMBY[10] campaigns that focus on particular local concerns.

Chief among local anxieties about proposed incinerators is their potential impact upon the health of the people who live in the vicinity. Public perceptions of incineration have been shaped by belated recognition that earlier generations of waste incinerators were highly polluting, perceptions that, paradoxically, are likely to have been reinforced by the dramatic tightening of regulations of incinerator emissions, especially of dioxins, since 1989. Several inspectors' decisions from 1994 onward considered that public concern and perception of risk, whether well-founded or not, is a material consideration in the determination of planning applications. Nevertheless, public concern about health impacts does not appear to have been the critical factor in any of the decisions concerning proposed incinerators. Friends of the Earth (2002) concluded its review of incinerator planning decisions with the advice that arguments invoking fears of health impacts were likely to be persuasive only if made in respect of particular sites rather than as concerns about incineration in general. Planning law has, however, been devised to ensure that real or alleged health implications of proposed developments are not considered as part of land-use planning decisions.

Certification that a proposed development will in all likelihood be compliant with pollution controls is the responsibility of the Environment Agency, and planning authorities are required to assume that the Agency will not issue a certificate for a potentially non-compliant facility. The Agency is an autonomous statutory authority whose mandate is to regulate environmental matters in accord with the best available scientific evidence. Its views concerning incinerators are unambiguous:

> Studies into the health of communities living near to incinerators have not found any convincing links between incinerator emissions and adverse effects on public health. We work with health authorities and the Health Protection Agency to investigate local concerns. We regulate all waste facilities, including energy from waste incinerators, to prevent or minimise any risks to the environment or health. Cuts in emissions over the past decade have greatly reduced any potential health risks. (Environment Agency 2009)

The Health Protection Agency (HPA 2005, p. 4) gave scarcely more ammunition to anti-incinerator campaigners:

> Incinerators emit pollutants into the environment but provided they comply with modern regulatory requirements, such as the Waste Incineration Directive, they should contribute little to the concentrations of monitored pollutants in ambient air. Epidemiological studies, and risk estimates based on estimated exposures, indicate that the emissions from such incinerators have little effect on health.[11]

In its response to the British Society for Ecological Medicine report, *The Health Effects of Waste Incinerators*, the HPA concluded that:

> As there is a body of scientific evidence strongly indicating that contemporary waste management practices including incineration, have at most, a minor effect on human health and the environment, there are no grounds for adopting the 'precautionary principle' to restrict the introduction of new incinerators. (HPA 2006, p. 1)

The problem is, however, that whatever risks there are, they are very unevenly distributed. Poorer communities suffer most from environmental pollution, and '50 per cent of operating municipal waste incinerators in England are located in the most deprived 10 per cent of wards' (FoE 2004). New waste incinerators are most often proposed for areas in which there is already a concentration of social and environmental deprivation that may be presumed to have negative effects upon human health. Ecological studies of the impact of municipal waste incinerators generally conclude that it is difficult to demonstrate any impacts upon the health of people living nearby. Those studies that do suggest higher rates of cancer or cardiac disease among the neighbours of municipal waste incinerators generally make the point that the apparent effects may be confounded by other factors, environmental or other, and that the observable increases in pathology are difficult or impossible to distinguish from the effects of other factors in urban environments (see Rootes, introduction to this volume).

Residents' objections that proposed new developments call upon them to bear more than their fair share of the environmental burdens of modernity are common in communities already affected by pollution.[12] Such arguments were especially fierce in Bexley (south-east London) where residents resisted proposals for a waste incinerator at Belvedere four times and through three public inquiries over a period of 15 years (1991–2006). Bexley already suffered severe air pollution from a concentration of upwind industrial activities and vehicle emissions. It particularly angered Bexley residents that the massive incinerator, finally approved after a protracted public inquiry over-ruled objections from the borough council, will principally incinerate waste transported downriver from more affluent west London.

Oppositional discourses highlighting environmental inequities and perceived threats to already impaired public health have, in the face of the derogation of responsibility for such matters to the Environment Agency and the HPA, had no discernible impact upon decisions concerning incinerators.

Universalist arguments are dismissed for want of compelling evidence of harm, and particularistic arguments are met with assurances that site-specific regulation and monitoring will keep emissions within safe limits. The last recourse of campaigners is therefore to attempt to demonstrate that, because of some peculiarity of local conditions, permitted limits cannot reliably be maintained or that the local environment is peculiarly susceptible to damage by even low levels of emissions.

Particularistic local campaigns often articulate issues of ecology that more universalistic campaigns overlook, but such particularistic concerns seldom persuade planners or political decision-makers. Indeed, preoccupations with the universalist demands of economic development have often overridden the legally protected status of designated areas of special scientific interest and/or special landscape value (as they did during the 1990s road-building programme). But if particularistic arguments fail to convince planners and decision-makers, they may nevertheless involve and encourage local people desperate for any weapon with which to defend their habitat. Because such arguments tend to invigorate and prolong campaigns, they make it less likely that those campaigns will be overlooked politically even where decision-makers find the substance of local concerns uncompelling. Developers faced with the prospect of protracted arguments with planners and regulators as they attempt to address issues raised by local campaigners must anticipate the cost of long delays before final decisions can be reached, and uncertainty about outcomes given the higher likelihood of political intervention in long-running disputes. Where opposition has been sustained, developers have often reassessed the commercial risks of the proposed projects and abandoned them. Thus intense NIMBY campaigns have sometimes succeeded even in the absence of good universalist arguments (Rootes 2006).

The adoption of universalistic discourse should, nevertheless, be advantageous to local campaigners because it legitimates their claims by reference to broader principles and because, by transcending the particularities of the locality, it facilitates the building of alliances with non-local actors who may be expected to bring additional resources to the campaign. This certainly appears to have been the case at Byker. There the opposition campaign started out by employing a confrontational style to stop the plant being built but quickly developed a broader strategic discourse by which it sought to brand incineration as bad and recycling as good, 'enrolling intensely local support in Byker [but also] attracting national and international contributors to [the] select committee process' by which it developed for the city an alternative waste strategy 'based on a shift from technologies of disposal to people-centred resource recovery' (Watson and Bulkeley 2005: 419–420).

Yet in other cases where sophisticated arguments and alternative waste management strategies have been proposed, the outcomes for campaigners have been less happy. The campaign against the incinerator proposed for

Allington, near Maidstone in Kent, primed by a year-long campaign at nearby Halling where the fiercely resisted proposal for a waste incinerator was suddenly dropped when the Allington site became 'commercially available', clearly articulated universalistic arguments as well as proposing practical alternatives such as community recycling and composting schemes as well as large-scale in-vessel composting. Such arguments and proposals had no visible impact upon a waste planning authority committed to sponsoring an incinerator, and campaigners were advised by local councillors who were themselves opposed to the incinerator that they should instead focus upon arguments specific to the site and especially upon its status in the local structure plan as a protected greenfield space integral to the 'green wedge' between two conurbations. Thus because universalistic considerations were considered to have no bearing on the land-use planning decision that would determine the issue, campaigners keen to advance alternative strategies for waste management were urged back to the essentially NIMBY ground of a particularistic defence of place (Rootes 2006). Moreover, because the waste authority, exercised by the imminent expiry of its contracts for the landfilling of waste, was bent on securing planning permission for the incinerator as quickly as possible, campaigners had little time in which to develop broad alliances beyond Kent.

Clearly there is more going on here than the competing claims of alternative discourses. Not only does the discourse employed by campaigners not appear to have been decisive in determining the outcomes of incinerator proposals but the discourse of campaigners has itself been shaped by the structure of political opportunities inscribed in the planning system. In order to influence the decisions concerning particular proposed waste facilities, universalistic arguments, however good, must be made in a local context. Because any proposal for a new waste facility is lodged in relation to a particular site, objectors must show that *that* site is not suitable for *that* purpose. The arguments they are thus required to make are necessarily particularistic because, ostensibly, the decision will be made on the basis of narrow local land-use planning considerations. Wider policy considerations may be involved, but they are substantially irrelevant to the land-use planning process. For that reason, objectors are advised to concentrate upon land-use planning issues and encouraged to become clever NIMBY campaigners rather than attempt to engage in a larger policy debate. Thus the utility to local campaigners of universalistic ecological discourse is severely limited by the structure of political opportunities (Rootes 2006).

Political opportunities structures

Political opportunities structures in the strict sense not only influence the strategy, tactics and discourse of campaigners but impact directly upon siting decisions and the outcomes of campaigns. Waste planning authorities in England are in general coterminous with first-tier local authorities: the county councils and the increasingly numerous unitary authorities that combine the functions of county and district councils. Whereas the counties are generally geographically extensive, the unitary authorities are relatively geographically compact. The distribution of new waste incinerators among them is striking. Of the 18 waste incinerators operational in August 2009, only five were entirely new and had commenced operation since 1998.[13] Of these five, four are within the boundaries of county councils, as was the fifth at the time when the incinerator was first proposed.[14] Three plants were under construction in 2009.[15] In the ten years and eight cases since 1998, there has been only one in which a unitary council has approved a new waste incinerator within its jurisdiction. Moreover, most of the sites currently proposed for new incinerators are within counties rather than geographically compact unitary authorities.

Whereas county councils have quickly granted approval for the siting within their own county boundaries of incinerators sponsored by the local waste planning authority, unitary authorities generally have not. Thus Kent, Hampshire and North East Lincolnshire County Councils approved waste incinerators but, in the new unitary authorities excised from Hampshire and Kent under local government reorganisation,[16] as in unitary Bexley, local councillors opposed incineration, and incinerator proposals were either abandoned[17] or approved only after the Secretary of State overruled the local authority following a public inquiry.[18]

If the new local authority boundaries were the final nails in the coffins of incinerators proposed for the Medway area, they ensured approval of an incinerator at Allington, on the very edge of the redrawn County of Kent, not least because most of the councillors who took the decisions represented voters living up to 40 miles away in east, west and north Kent. Many of these councillors were being strongly lobbied by local campaigners and the CPRE against landfill operations in their own areas, and so had no interest in rejecting an incinerator that was represented to them as a solution to the increasingly pressing problem of finding alternatives to landfill, especially because any undesirable effects of the incinerator would not be felt by their supporters. A similar political logic applied in Hampshire (Rootes 2001).[19]

Thus the political opportunity structures inscribed in the political geography of local authority boundaries explain much of the pattern of new incinerator sitings. In geographically large counties such as Kent and Hampshire, it was much easier to get a majority of councillors to take locally unpopular decisions than it was in the geographically compact unitary authorities, Medway and Portsmouth, in which councillors, having 'no place to hide' from unpopular decisions,[20] rejected incineration as a means of waste disposal.

This pattern has been repeated in the subsequent decade, but with the added twist that unitary authorities, rather than risk unpopularity by choosing urban sites within their boundaries, have tended to contract with neighbouring counties to locate waste incinerators beyond the boundaries of the unitary authorities.[21]

Not all counties have, however, been so open to proposals for incinerators. Lancashire County Council, for example, despite formulating an early waste strategy that envisaged a major role for incineration, in its 2004 strategic review recognised that changes in government and European policy and technological developments made incineration a much less attractive option and instead resolved to focus upon forms of waste management that did not entail incineration.

Moreover, some unitary authorities *have* embraced incineration.[22] However, in all but one of these cases incineration had been employed as a means of waste disposal before the advent of modern energy-from-waste (EfW) incinerators.[23] This does not, however, mean that the refitting of old incinerators or their replacement with new EfW plants has everywhere been uncontroversial. There was bitter opposition to Sheffield City Council's proposal, narrowly approved in 2002, for a new incinerator to replace one identified before its closure in 2001 as the most polluting in Britain. The Byker ash scandal and the concerted campaign of opposition to the construction of a replacement incinerator at Byker was instrumental in persuading Newcastle City Council to adopt a 'no incineration' policy for its waste. More recently, proposals to expand the existing Nottingham incinerator have encountered stiff local resistance.

If local campaigns have made life difficult for waste planners in unitary authorities,[24] campaigners have also sometimes been successful in thwarting County Councils' plans for waste incineration. Most conspicuously, Surrey County Council, faced by determined opposition campaigns, failed to gain approval for waste incinerators on any of its three preferred sites.[25] West Sussex County Council in 2002 deferred to vociferous local protests and abandoned plans for a waste incinerator.

The Surrey and West Sussex cases remain exceptional, however.[26] Most of the failed incinerator proposals in the shire counties have been initiatives of waste management companies rather than projects endorsed by the county authorities. Thus Kent County Council, having approved and committed itself to the Allington incinerator, saw no need for a second incinerator that might make the one at Allington less commercially viable, and so refused permission for a waste incinerator at Ridham in 1999 (Rootes 2006).[27] Especially in light of increasing focus upon recycling, the existence or approval of a waste incinerator in a region makes it increasingly difficult for developers to demonstrate the local need for increased incinerator capacity.[28]

Thus although political structures differentiate outcomes of campaigns in densely populated, geographically compact unitary authorities from those in geographically extensive counties, they are unable to explain differences in outcomes *within* those counties. In these cases it is apparent that it is the sequence of events and the changing policy context that has had most influence on the results of campaigns.

Changing policies, changing practices

From 2001, in the wake of the Byker ash scandal, national political and environmental groups became more willing to take up the incineration issue.[29] During 2001–2002, Greenpeace, which had not previously been active on municipal waste issues in Britain, conducted its 'Incinerator Busters' campaign with characteristically spectacular protests against incinerators in Sheffield and London.[30] The protests were designed to highlight health risks posed by the emission of dioxins, all three incinerators having been at least briefly in breach of permitted emission levels, and to encourage consideration of alternative modes of waste disposal such as recycling and composting. In response, the opposition Conservative Party called for a moratorium on the building of new waste incinerators, its environment spokesman demanded that government investigate Greenpeace's claims about dioxin emissions, and Liberal Democrat Councillors and Green Party members of the London Assembly supported the Greenpeace actions. The political context was changing and contention over waste management was shifting from battles over siting decisions to debates about sustainable strategies that bring issues of consumption and civic responsibility to centre stage.

These changes were reflected in the recommendations of the UK government's Strategy Unit (2002) in its review of waste policy. Incineration was not ruled out but recycling targets of above 45% were recommended, and waste authorities were cautioned about the disadvantages of becoming locked into long-term contracts with incinerators. Government, responding to the Strategy Unit's recommendation of measures to encourage waste reduction and recycling, has employed three principal policy instruments: progressive increases in the landfill tax, a Landfill Allowance Trading Scheme, and statutory recycling targets.

The Landfill Tax was introduced in 1996 by the then Conservative government. The first explicit eco-tax in the UK, it was designed to make the cost of disposing of waste to landfill better reflect its environmental impact. Given greater urgency by the EC Landfill Directive, it is a key mechanism to enable the UK to meet its obligations to reduce the landfilling of biodegradable waste. By progressively increasing the cost of landfill, it was intended to provide financial incentives for the development of other, less environmentally damaging forms of waste disposal, including more advanced waste technologies whose gate fees were, without the tax, higher than those for landfill (DEFRA 2007a). Although the tax is levied on all waste sent to landfill, the rate for inert waste (principally construction/demolition waste such as concrete, brick and glass as well as uncontaminated soil, clay and gravel) has remained low (£2.50 per tonne in 2008). However, the tax on 'active' waste (waste that gives off emissions) has steadily escalated. From 1999 to 2004, the 'landfill tax escalator' was £1 per tonne, increasing to £3 per tonne from 2005. The tax had reached £24 by 2007 when the government announced that from April 2008 it would increase by £8 per annum. The escalator was extended in

2009 and the landfill tax will, as a result, reach £72 per tonne by 2013.[31] Thus the screw has been tightened progressively and with increasing force. Waste authorities and developers responded, tentatively at first, but with a growing sense of urgency.

The Landfill Allowance Trading Scheme (LATS) was introduced in 2005 as a device to ratchet down the amount of waste consigned to landfill. Each waste disposal authority (WDA) is given the right to landfill a certain amount of biodegradable municipal waste each year, allowances being allocated at a level that will enable the UK to meet its targets under the Landfill Directive. WDAs may trade allowances with other authorities, save them for future years, or use some of their future allowances in advance to enable them to plan investment in alternative means of waste disposal. The allocation reduces progressively year on year until 2020, and a fixed penalty of £150 per tonne is incurred if a WDA fails to hold sufficient allowances for the amount of biodegradable waste it sends to landfill in a scheme year (DEFRA 2009). Because they can sell their surplus allowances to generate income, WDAs are thus encouraged to divert more waste from landfill and to promote waste minimisation, reuse, composting, recycling and recovery.

The financial implications of the sharply increasing costs of landfill to waste disposal authorities have transformed a growing sense of urgency into something like panic. The consequence has, in the years since 2005, been an unprecedented rash of proposals for incinerators, anaerobic digesters and gasification plants of various kinds.

The increased emphasis upon recycling in EU and national waste strategies has outdated local waste plans formulated less than a decade earlier and has raised increasing doubts about the viability of waste strategies to which incineration is central. The sustained rise in recycling rates since 2002 – to an average 37% in 2008 – promises to reverse the conditions that made incineration appear attractive in the first place. In 1990, the problem was that waste arisings were growing at an estimated 3% p.a. but by 2008 they were *falling* at 3% p.a. (DEFRA/National Statistics 2009). As recycling and composting become more popularly entrenched, so local waste authorities face the prospect of declining waste arisings, especially of readily combustible paper, card, plastic and organic material. In such circumstances, the disadvantages of long-term contracts to deliver combustible waste to incinerator operators are considerable.

The waste industry is, however, increasingly rationalised, and the multi-national waste management companies that now dominate it tend to have preferences for technologies of which they already have experience. Thus there is an inherent bias toward older technologies that have proven track records and reliably estimable costs. It is generally the upstarts in this increasingly concentrated industry that propose new, experimental technologies, or that propose to import technologies known to work elsewhere but untried under local conditions. Waste management companies that already have long-term contracts to dispose of municipal waste have, as a result, often continued to propose mass burn incineration.

There is some irony in the fact that many councils should be continuing to consider new large-scale waste incinerators when the case of the most recently commissioned waste-to-energy plant is so discouraging. Not only did the Allington plant take more than ten years from planning application to commercial operation but, in the meantime, the supply of combustible waste has diminished. As a result, incineration has become an extremely expensive operation for the waste authority, which is locked into a contract to supply 500,000 tonnes of waste per annum. Kent County Council (KCC) admits that Allington has been a costly mistake,[32] and in order to achieve the tonnage of waste required, in addition to waste from the four adjoining boroughs that Allington was designed to serve, from 2009 waste will be trucked in from Surrey and Thanet, more than 50 miles away.

Yet although EC and government policy continues to encourage recycling, it is not certain that the quantities of waste requiring treatment will continue to decline. Local governments are under increasing financial pressure, especially with the parlous state of public finances since the financial crisis of 2008–2009. As a result, waste authorities are increasingly concerned to seek the lowest cost option, even where it means compromising their recycling targets. Thus Ashford Borough Council (Kent) announced in March 2009 that it would end its scheme for the collection and composting of green waste and instead accept green waste alongside its other municipal waste sent for incineration at Allington. Other waste collection authorities, similarly alarmed by the costs of collecting recyclables, would probably follow suit were they not locked into contracts for the collection of recyclables with waste management companies.[33]

There may, however, be yet another twist in policy that will tip the balance against incineration. Under existing UK legislation, bottom ash and slag from waste incineration have been treated as 'inert', but under EU legislation they are not and so would be liable to the standard rate of landfill tax. The UK government in 2009 embarked on a consultation exercise to consider the implications of reclassifying these as active wastes. 'Modernising the definitions used for landfill tax to reflect environmental protection legislation and waste industry developments would help ensure that the landfill tax has its intended effects in providing an incentive to find more sustainable uses and treatments for waste. Restricting wastes which can qualify for the lower rate of tax would strengthen the environmental position of the tax because only those wastes which were inert based on an up-to-date understanding of their impact would be lower rated' (HM Revenue and Customs 2009, p. 40).

One implication of so doing would be significantly to increase the costs of waste incineration, particularly those forms that produce relatively large quantities of residual ash.

Climate change and waste policy

Since 1990, the recovery of energy from waste has been seen by both government and opposition as a potential contributor to reducing the UK's overall carbon emissions by substituting for fossil fuels presently used to generate electricity (Gandy 1994, pp. 67–68). Thus for two decades incineration of waste without energy recovery was ranked below incineration *with* energy recovery, but there was little analysis of the contribution of the latter either to electricity generation or, more particularly, to reducing carbon emissions. Hogg (2006) concluded that, in terms of overall carbon emissions, waste incineration only performed better than combustion of fossil fuels where the heat produced by incineration was also captured, as in modern combined heat and power (CHP) facilities. However, for CHP to be economic, the incinerator needs to be located very close to potential markets for its heat. The difficulty of finding suitable sites in urban areas, and the general perception, rooted in past experience, of incinerators as bad neighbours, encourages developers and waste authorities to seek sites remote from neighbours. Most waste incinerators in the UK, even where they are sited in urban locations where a ready market for the heat might be accessible, do not capture the heat. In terms of its efficiency in producing electricity – and hence abating the costs of waste disposal – incineration compares poorly with gasification; biogas, which requires considerably less capital investment, produces more energy at lower cost (Murphy and McKeogh 2004).

The UK Energy White Paper (DTI 2007) recognised the energy policy benefits of recovering energy from residual waste. In particular, waste biomass is seen as a renewable energy source, and biogas as well as electricity produced from the biomass content of waste treated in gasification, pyrolysis, anaerobic digestion, and good quality CHP plants were supported under the Renewables Obligation. In line with EU policy, incinerators without CHP are considered to be primarily waste disposal plants, and the electricity they produce is not classed as renewable energy. Although government expressed no general preference for one energy recovery technology over another, it singled out anaerobic digestion for special encouragement on the grounds that recent research showed that it has 'significant carbon and energy benefits over other options for managing food waste' (DEFRA 2007b). The UK Renewable Energy Strategy reaffirms government support for anaerobic digestion (DECC 2009, pp. 109–110).

When Hogg *et al.* (2008) systematically compared and ranked various waste treatment technologies in terms of their overall contribution to carbon reduction goals, incineration – with or without CHP – performed poorly, especially by comparison with anaerobic digestion. Even with CHP, incineration ranked only 19th out of 24 alternatives in terms of measurable externalities, with incineration producing only electricity (the form of EfW plant most common in Britain) a lowly 22nd. Incineration produces lower greenhouse gas emissions than landfill but more than twice as much as plasma gasification or anaerobic digestion. Making reasonable assumptions about likely changes in the waste stream (relatively more plastic, less compostable green waste), incineration with energy recovery becomes more than three times as expensive, externalities included, as the best available alternatives.

A further advantage of anaerobic digestion and, indeed, newer gasification technologies,[34] is that they are modular and do not necessarily entail the large-scale plant required to make conventional waste incineration economic. Large-scale incinerators, especially where they are sited in remote rural locations in order to minimise nuisance to neighbours, require large numbers of vehicle movements that are themselves a major source of pollution and of nuisance. This is a major consideration because recent research concludes that centralised, large-scale facilities have greater overall negative environmental impacts than small-scale, distributed systems. Centralised waste schemes generate more traffic, incur higher fuel costs, and emit more carbon per tonne of waste processed (Bastin and Longden 2009).[35] Indeed, the scale of waste management facilities appears to be more significant than the choice of technology (Longden *et al.* 2007).[36]

Taking these three considerations – capital cost, technology-specific carbon emissions, and scale and location of facility – together, and supposing that they are not overridden by short-term financial considerations, it would appear that, in most circumstances, large-scale waste incineration may indeed be a 'dying technology' (Tangri 2003).

Success at last?

If incineration does play a restricted role in waste management in England in future, will that count as a success for the many anti-incinerator campaigns? There can be no doubt that, besides drawing attention to the unsuitability of particular proposals for particular sites, local campaigns have, more generally, focused attention upon waste and the means of disposing of it. They may not have won arguments about the health impacts of incineration, but they have established that genuine public concern about possible health risks is a material consideration for waste planners and an obstacle that waste management companies proposing incineration have, on each occasion, to confront. They have raised questions of equity and environmental justice by highlighting the extent to which waste incinerators were concentrated in socially deprived areas. They have highlighted concerns about the environmental impacts of incineration that have spurred commitment to the minimisation of emissions, and they have posed questions about the impacts of large-scale waste treatment facilities that have encouraged planners to take the proximity principle more seriously. In so doing, they have compelled waste planners at all levels to take the alternatives to incineration more seriously than they otherwise might.

Campaigners may not have won the argument about the health impacts of incineration, but they have been proved right in many of their other objections – the achievability of high rates of recycling, effects of long-term contracts, the viability of alternatives – and their efforts have contributed to the re-evaluation of waste management strategies by national government and local waste authorities, and to reconsideration of the economics of incineration by the waste management companies.

Anti-incineration campaigners have not, of course, succeeded in preventing the construction of every proposed incinerator, but they have very often elicited changes in design or associated transport arrangements or compensatory environmental improvements that have softened the blow even in defeat. Even where they have in particular cases failed, anti-incinerator campaigners, by obstructing and delaying proposals, have bought time for the marshalling of new evidence against incineration and, more importantly, for the development of new waste technologies, and for proper evaluation of the efficiency of incineration in achieving policy objectives in respect of recycling, reduction of waste consigned to landfill, energy generation and lowering emissions of greenhouse gases. Because these policies have themselves been changing, it can be counted as a success for the campaigners that they have helped to limit the extent to which the UK is locked into technologies that are now shown to be sub-optimal from the point of view of public policy objectives.

Of course, none of these successes is simply the achievement of bands of local heroes. If national environmental NGOs have played a limited role, FoE in particular has been a fairly continuous resource for local campaigners, providing advice on how to campaign and what issues to emphasise and how, and it has sought to empower local campaigners by fostering networking among them and encouraging them to recognise the expertise they themselves possess.[37] FoE has, moreover, played an important role in lobbying at national level for recycling and against incineration, and, by commissioning research into the climate change impacts of competing waste technologies (Hogg 2006), it has encouraged government to embrace anaerobic digestion as the most appropriate technology for the treatment of biodegradable waste.

As ever, the impacts of campaigners are rarely direct and unmediated. More usually, they are mediated through the discourses and practices of politicians, planners, the judiciary and the waste management companies themselves. The last two decades have seen a remarkable sequence of contention the outcome of which has not simply been victory for anti-incineration campaigners but significant progress toward the development of a regime and infrastructure of waste management that is much more sophisticated than that which existed in 1990.

Conclusion

The high profile and effective networking of anti-roads protests in Britain in the 1990s raised expectations that something similar might be achieved by campaigns against waste incineration. It never happened. Whereas road-building was a central policy to which Conservative governments stuck doggedly through more than six years, the Labour government appeared only briefly, in 2000, to favour incineration; it quickly distanced itself as soon as controversy erupted in 2001. Thus waste incineration was never a core policy for Labour as roads were for the Conservatives, it only very briefly became a national issue, and political allies were correspondingly harder to find. Once it became clear that there would be no strong central government push for waste incineration, Greenpeace, which had joined the battle in 2000, beat a strategic retreat. FoE maintained an interest in waste management policy and continued to support local communities, but until very recently there were too few local campaigns at similar stages of development to sustain a national network.

If political opportunities in the general sense explain the relatively low profile of anti-incinerator campaigns, political opportunity *structures* in the strict sense (see Gamson and Meyer 1996, Rootes 1997, 1999) explain the fragmentation of anti-incinerator campaigning and much of the variation in the outcomes of local campaigns. The planning system compels local campaigners to frame their objections in particularistic terms, and the structures of local government make it more or less likely that councils will adopt incineration as a waste strategy or approve particular incinerator proposals, geographically extensive county councils being more likely than compact unitary authorities to approve incinerators. However, time, timing and especially changes in government policy have changed the parameters within which contention over waste is conducted. Recycling has rapidly increased as the government's policy instruments have begun to bite. As climate change has risen up the agenda, and as it has been recognised that waste incineration may not, after all, be as climate friendly as had previously been claimed, so there has been renewed interest in alternatives such as anaerobic digestion. Technological developments have brought the promise of new, smaller-scale means of waste disposal that minimise adverse environmental impacts.

It is, then, somewhat ironic that an autonomous network of anti-incinerator campaigners should be brought into existence at the very time that incineration's appeal as a means of waste management is waning. Last-ditch battles are, however, often the bloodiest, and for local campaigners the struggle is not yet over.

Acknowledgements

Research reported here was in part supported by the European Commission DG XII – Science, Research and Development (*Policy Making and Environmental Movements: Comparative Research on Waste Management in the United Kingdom and Spain*, coordinated by Enrique Laraña, Universidad Complutense de Madrid, and *The Transformation of Environmental Activism*). The author is indebted to Michelle Asbury, Rebecca Whale, Richard Steer, Clare Saunders and Gabriel Swain for research assistance. Thanks, too, to Anna Watson, Neil Carter and Chris Pickvance for helpful comments on the penultimate draft.

Notes

1. Waste to energy is not new in Britain. The first operational incinerator, or 'destructor', began operation in 1877. By 1912, there were over 338 refuse incinerators in Britain, of which more than 80 also generated electricity (Clark 2003).

2. This account is based on an interview with Greenpeace campaigner, Mark Strutt (Wood 2006).
3. Groups represented included those from Hull, Bradford, Essex, Derbyshire, Brighton, Kent, London, Milton Keynes, Guildford and Wales (http://www.greenpeace.org.uk/media/press-releases/activists-halt-construction-of-new-incinerator, accessed 15 March 2009).
4. This account is based on an interview with Greenpeace campaigner, Mark Strutt (Wood 2006).
5. FoE produced its first guide for anti-incinerator campaigners in 1997 but before that was supporting communities at local planning inquiries.

> Many of FOE's local groups have campaigned against incineration for many years and we supported them with funds, advice, resources etc. The campaigners had a chance to sit on an advisory group for the waste team to help us ensure we were providing them with the support they needed and that they understood how our national campaigning would help influence what they were facing on the ground. Therefore we campaigned on the new waste planning guidance PPS10 to ensure that cumulative impact assessments have to be done by local authorities so that the same communities are not continually faced with waste management infrastructure. We also ensured by our lobbying of what was ODPM [Office of the Deputy Prime Minister] at the time that health would still be a material consideration as this was such an important issue for the community groups. (Anna Watson, email to author, 11 September 2009)

It appears that the decision to produce its guide for anti-incinerator campaigners was taken because it was considered that supporting local campaigners at public inquiries was too demanding of FoE's limited resources, resources that might be better deployed in building capacity in the communities themselves.

6. UKWIN's 45 members in August 2009 included FoE and 17 local FoE groups, and of the 33 anti-incinerator campaign groups listed in updates on UKWIN's website 2009, four were local FoE groups.
7. Anna Watson, who worked at FoE with community groups opposing incineration from 2003 to 2008, did all the legwork, collecting material on the various options of what a network might look like and how it might work. FoE produced a report (Wood 2006) as background for the initial meeting, which was attended by 50 people from 30 groups. The meeting agreed to set up a steering group, with Anna Watson as a member. Anna then worked to sound out potential funders and to pre-arrange funding for a part-time organiser, the money coming from a private charitable trust, the Ecology Trust, supported by a donation from the Manuka Club (Anna Watson, interviewed by author, 11 September 2009).
8. 'We used to say to them, if that's something you are worried about, go ahead and ask the questions, but be armed (and we would provide them with information), but do not expect to win on those issues because they [the planners] have got such a tight line on that ... If you're worried about the health impacts of incinerators, go ahead and say so, but we'd say there are probably bigger health issues in your community that you should be worried about – traffic, for example' (Anna Watson, interviewed by author, 11 September 2009). Greenpeace produced a quasi-academic booklet summarising the scientific evidence concerning incineration and health, and distributed it to local groups and activists, but it was up to them how they used it, and 'as local campaigners weren't Greenpeace-branded, then if there were exaggerated claims then they weren't challenged' (from interview with Greenpeace campaigner, Mark Strutt, in Wood 2006).

9. The ecological modernity of waste incineration, even with energy recovery, has however been sharply challenged by FoE and by the waste management company BIFFA for whom it is simply an end-of-pipe solution that does not tackle the fundamental problems of waste minimisation through better product design. BIFFA contended that a large-scale incinerator building programme would inhibit the development of greener technologies (BIFFA 1999, p. 10).
10. My use of the term 'NIMBY' ('not in my back yard') is intended to be simply descriptive and does not carry the stigma of hypocritical selfishness usually associated with it (see Rootes 2007).
11. Its 2009 update reaches essentially the same conclusion (HPA 2009).
12. See, e.g., Schlüter *et al.* (2004): residents of Grangemouth, the site of a notable concentration of petrochemical plants, when confronted with proposals for expansion of those facilities objected that 'enough is enough'. Similar sentiments were voiced in 1996 when waste incinerators were proposed at Halling and Kingsnorth in the Medway valley, whose residents complained of existing poor air quality, and in 1999 in Slough where a new waste incinerator (Colnbrook) was approved for a town that already had a hazardous waste incinerator and a large combined heat and power plant burning bio-mass and refuse-derived fuels.
13. Chineham and Marchwood in Hampshire, Grimsby in North East Lincolnshire, Allington in Kent, and Portsmouth. While the incinerators at Grimsby and Allington had no local precedents, the EfW plant at Chineham, commissioned in 2003, was built on the site of an old incinerator closed in 1996, and that at Marchwood, commissioned in 2007, is close to the site of an old incinerator closed in 1996. Hampshire was unusual among county councils in the extent to which it incinerated its waste; until 1991 it had a network of six municipal waste incinerators burning 30% of Hampshire's waste, as compared with the then national average of 7% (Petts 1995, p. 523).
14. Portsmouth was excised from Hampshire to become a unitary authority in 1998, and its new council voted, in response to a concerted local campaign, but against the advice of council officers, to reject the proposed incinerator. There had previously been an incinerator on the site but this closed in 1991 due to the costs of meeting the environmental standards established by the Environmental Protection Act 1990. The proposed incinerator was much larger than the old and was designed to serve a larger geographical area.
15. Belvedere, approved only after the local borough council, which had refused planning permission, was overruled by the Secretary of State; Newhaven, approved by East Sussex County Council; and Colnbrook, approved by Slough unitary authority. Slough is only at first glance an anomaly. Slough already had a long history of waste incineration, the operator of the proposed incinerator was a locally owned company and gained planning permission in 1999. The incinerator was, nevertheless, vigorously but very belatedly – at the IPCC stage – opposed on health grounds by a campaign group (SAIN).
16. Portsmouth in 1997 and Medway in 1998, respectively.
17. At Halling and Kingsnorth, both in Medway.
18. As at Portsmouth and Bexley.
19. And later in East Sussex, where opponents of landfill supported proposals for an incinerator elsewhere in the county (note, e.g., the campaign by BALI – against landfill at Bexhill).
20. Officer, Portsmouth City Council, interviewed by author, 14 February 2000.
21. Thus York City Council (unitary) promised that there would be no incineration in York, but contracted with North Yorkshire County Council to find a site for a possible incinerator in North Yorkshire to serve the needs of York. Councillors in Brighton and Hove (unitary), adamant that there could be no

suitable site there even while insisting incineration posed no threat to health (Baker 2003), contracted to incinerate Brighton's waste at Newhaven in the neighbouring county, East Sussex. In Hull, where the local unitary favoured incineration but was forced by public opposition to abandon an urban site, it has collaborated with neighbouring East Riding of Yorkshire County Council to select a site straddling the boundary between Hull and East Riding.

22. They include Birmingham, Nottingham, Kirklees, and Teesside

23. The exception – SELCHP – was a collaborative venture between three south-east London boroughs that lacked landfill capacity. Designed as a state-of-the-art combined heat and power plant, it was approved in 1989 at a time when waste was viewed as a substitute for fossil fuels. Although it has been rated as Britain's least polluting waste incinerator, despite its urban location it has never found a market for the heat it produces.

24. In Bristol in 2009, the Labour-led council's plans for a large and hotly opposed waste incinerator were held responsible for its loss of local elections to the Liberal Democrats.

25. At Capel, Redhill and Guildford. At Guildford, the intensity of the campaign against the Conservative-dominated council was even credited with securing the election in 2001 of an anti-incineration Liberal Democrat MP.

26. Essex presents the interesting case where a county council wishing at least to keep the incineration option open has for a decade been thwarted by concerted action by a consortium of district councils opposed to incineration. This may be explained by political divisions between the Conservative-controlled county and the anti-incineration Labour and Liberal Democrat majorities on urban district councils, but the early and sustained involvement and lobbying of Friends of the Earth since the waste plan was first being formulated appears to have been critical (see Whitney 2000, Rootes 2001).

27. Ironically, county councillors, in order to justify their decision in respect of Ridham, deployed many of the arguments made only a year or two earlier by opponents of the Allington incinerator (Rootes 2006).

28. At Kidderminster, Worcestershire County Council in 2001 refused an application to build a waste incinerator that would have been the sixth in the West Midlands region.

29. See Ares and Bolton (2002) for a useful summary of developments to that date.

30. Edmonton in north London and SELCHP in south-east London.

31. The Conservative opposition environment spokesman announced in June 2009 that, in order to give the waste industry the certainty necessary to investment, a future Conservative government will maintain the tax at a minimum £72 per tonne until 2020 and that thereafter it will be indexed in line with inflation.

32. In August 2008, KCC revealed that the contract under which it is to supply the incinerator operator with waste will cost it £1 million a year for 25 years. The contract, negotiated without public scrutiny ten years earlier when the UK's recycling rate was under 10% and when recyclate had little value, treated waste as a disposal cost rather than a potential source of value. By 2008, Kent's recycling rate was over 32%, the economic value of recyclate had soared even as its availability increased, and the success of kerbside collections of paper for recycling and green waste for composting reduced the amount of combustible waste available for incineration.

33. To discourage such diversion to incineration of waste that might more appropriately be treated by anaerobic digestion, the government was in 2009 consulting about the possible introduction of a tax on incineration.

34. Whilst various gasification technologies appear to have advantages over direct incineration of waste in terms of carbon emissions and efficiency, they are not a panacea and perform less well than anaerobic digestion (FoE 2009, McKenna 2009).
35. Based on modelling of waste disposal options for Warwickshire and Cornwall.
36. Smaller facilities more easily gain planning permission. Thus a 120,000 tonne p.a. gasification plant in Doncaster was, in 2009, approved after only five months (Recycling & Waste World 2009). Smaller facilities are also more likely to find ready markets for the heat they produce, and so achieve optimal overall efficiency. The relatively small (56,000 tonnes p.a. capacity) Grimsby Efw/CHP plant supplies heat and power to an adjacent industrial estate. The larger Marchwood incinerator (165,000 tonnes p.a.) supplies heat and power to Southampton docks, but generates more heat than is needed by the docks, and so was in 2009 seeking an alternative market.
37. Anna Watson, interviewed by author, 11 September 2009.

References

Ares, E. and Bolton, P., 2002. *Waste Incineration,* Research Paper 02/34. London: House of Commons Library.

Baker, N., 2003. Summary of proof of evidence in respect of the Public Inquiry into the East Sussex and Brighton and Hove Waste Plan. Available at http://www.normanbaker.org.uk/local/incin.htm [Accessed August 2009].

Bastin, L. and Longden, D.M., 2009. Comparing transport emissions and impacts for energy recovery from domestic waste (EfW): centralised and distributed disposal options for two UK counties. *Computers, Environment and Urban Systems.* Available at http://www.sciencedirect.com/science?_ob=ArticleURL&_udi=B6V9K-4WJ918J-1&_user=10&_rdoc=1&_fmt=&_orig=search&_sort=d&_docanchor=&view=c&_searchStrId=1079439392&_rerunOrigin=google&_acct=C000050221&_version=1&_urlVersion=0&_userid=10&md5=6a2e8ddf1243a2c17d9bfeafda03da12 [Accessed 17 June 2009].

BIFFA, 1999. *A question of balance.* High Wycombe: BIFFA Waste Services.

Clark, J.F.M., 2003. *The burning issue: historical reflections on municipal waste incineration.* Stirling and St Andrews: AHRB Research Centre for Environmental History.

Cole, L.W. and Foster, S.R., 2001. *From the ground up: environmental racism and the rise of the environmental justice movement.* New York and London: New York University Press.

Dalton, R.J., 2008. *Citizen politics.* 5th ed. Washington, DC: CQ Press.

DECC (Department of Energy and Climate Change), 2009. The UK renewable energy strategy, Cm 7686. London: DECC.

DEFRA (Department of Environment, Food and Regional Affairs), 2007a. Waste strategy factsheets. Landfill tax. Available at http://www.defra.gov.uk/environment/waste/strategy/factsheets/landfilltax.htm [Accessed August 2009].

DEFRA, 2007b. Waste strategy factsheets. Energy from waste and anaerobic digestion. Available at http://www.defra.gov.uk/environment/waste/strategy/factsheets/energy.htm [Accessed August 2009].

DEFRA, 2009. Landfill allowance trading scheme. Available at http://www.defra.gov.uk/ENVIRONMENT/waste/localauth/lats/intro.htm [Accessed 15 September 2009].

DEFRA/National Statistics, 2009. Municipal waste management statistics, 6 August 2009. Available at http://www.defra.gov.uk/environment/statistics/wastats/bulletin09qtr.htm [Accessed August 2009].

DETR (Department of the Environment, Transport and the Regions), 1999. *A Way with waste: a draft waste strategy for England and Wales.* London: DETR.

DETR (Department of the Environment, Transport and the Regions), 2000. *Waste Strategy 2000 for England and Wales.* London: DETR.

Dodds, L. and Hopwood, B., 2006. BAN waste, environmental justice and citizen participation in policy setting. *Local Environment,* 11 (3), 269–286.

Dudley, G. and Richardson, J.J., 2001. *Why does policy change? Lessons from British transport policy 1945–99.* London: Routledge.

Environment Agency, 2009. Municipal waste incineration. Available at http://www.environment-agency.gov.uk/research/library/data/34419.aspx [Accessed 14 August 2009].

FoE (Friends of the Earth), 2002. Incinerator inquiries: a review of recent public inquiries considering planning permission for municipal waste incinerators. Available at http://www.foe.co.uk/resource/briefings/incinerator_inquiries.pdf [Accessed 26 August 2009].

FoE (Friends of the Earth), 2004. *Incinerators and deprivation,* Briefing. London: FoE.

FoE (Friends of the Earth), 2009. *Pyrolysis, gasification and plasma.* Briefing, January. London: FoE.

Gamson, W.A. and Meyer, D.S., 1996. Framing political opportunity. *In*: D. McAdam, J.D. McCarthy, and M.N. Zald, eds. *Comparative perspectives on social movements.* Cambridge University Press, 275–290.

Gandy, M., 1994. *Recycling and the politics of urban waste.* London: Earthscan.

Hajer, M.A., 1995. *The politics of environmental discourse: ecological modernization and the policy process.* Oxford: Clarendon Press.

HM Revenue and Customs, 2009. *Modernising landfill tax legislation.* London: HM Treasury.

Hogg, D., 2006. *A changing climate for energy from waste? Final report for Friends of the Earth.* Bristol: Eunomia Research & Consulting Ltd. Available at http://www.foe.co.uk/ [Accessed September 2009].

Hogg, D., *et al.*, 2008. *Greenhouse gas balances of waste: management scenarios.* Report for the Greater London Authority. Bristol: Eunomia Research & Consulting Ltd.

HPA (Health Protection Authority), 2005. *Municipal solid waste incineration.* London: Health Protection Authority.

HPA (Health Protection Authority), 2006. *Response to the British Society for Ecological Medicine report 'The Health Effects of Waste Incinerators'.* London: Health Protection Authority.

HPA (Health Protection Authority), 2009. *The impact on health of emissions to air from municipal waste incinerators.* London: Health Protection Authority.

Longden, D., *et al.*, 2007. Distributed or centralised energy-from-waste policy? Implications of technology and scale at municipal level. *Energy Policy,* 35 (4), 2622–2634.

McKenna, P., 2009. Could your trashcan solve the energy crisis? *New Scientist,* 2705, 33–35.

Milanez, B. and Bührs, T., 2007. Marrying strands of ecological modernisation. *Environmental Politics,* 16 (4), 565–583.

Murphy, J.D. and McKeogh, E., 2004. Technical, economic and environmental analysis of energy production from municipal solid waste. *Renewable Energy,* 29, 1043–1057.

Petts, J., 1995. Waste management strategy development: a case-study of community involvement and consensus-building in Hampshire. *Journal of Environment, Planning and Management,* 38 (4), 519–536.

Recycling & Waste World, 2009. Doncaster EfW plant approved in five months. *Recycling & Waste World,* 678 (20–26 August), 2.

Rootes, C., 1997. Shaping collective action. *In*: R. Edmondson, ed. *The political context of collective action*. London and New York: Routledge, 81–104.

Rootes, C., 1999. 'Political opportunity structures': promise, problems and prospects. *La Lettre de la Maison Française d'Oxford*, 10, 75–97. Available at: http://www.kent.ac.uk/sspssr/staff/academic/rootes.html [Accessed 17 October 2009].

Rootes, C., 2001. Discourse, opportunity or structure? Determining outcomes of local mobilizations against waste incinerators in England. *In*: *Paper presented to the workshop on 'Local environmental politics' at the 29th Joint Sessions of the European Consortium for Political Research*, Grenoble, 6–11 April.

Rootes, C., 2006. Explaining the outcomes of campaigns against waste incinerators in England: community, ecology, political opportunities and policy contexts. *Research in Urban Policy*, Vol. 10 (special issue on *Community and Ecology*, ed. by A. McCright and T. Nichols Clark), 179–198.

Rootes, C., 2007. Acting locally: the character, contexts and significance of local environmental mobilizations. *Environmental Politics*, 16 (5), 722–741.

Schlüter, A., Phillimore, P., and Moffatt, S., 2004. Enough is enough: emerging 'self-help' environmentalism in a petrochemical town. *Environmental Politics*, 13 (4), 715–733.

Strategy Unit, 2002. *Waste not, want not: a strategy for tackling the waste problem in England*. London: Cabinet Office.

Tangri, N., 2003. *Waste incineration: a dying technology*. Quezon City and Berkeley, CA: GAIA.

Walsh, E., Warland, R., and Clayton Smith, D., 1993. Backyard NIMBYS and incinerator sitings: implications for social movement theory. *Social Problems*, 40 (1), 25–38.

Walsh, E., Warland, R., and Clayton Smith, D., 1997. *Don't burn it here: grassroots challenges to trash incineration*. University Park, PA: Penn State University Press.

Watson, M. and Bulkeley, H., 2005. Just waste? municipal waste management and the politics of environmental justice. *Local Environment*, 10 (4), 411–426.

Whitney, P., 2000. The Essex waste incineration battle. *In*: *How to win: campaign against incinerators*. London: Friends of the Earth, 36–38.

Wood, A., 2006. *Campaign networking in six organisations*. A report for Friends of the Earth. London: FoE.

A burning issue? Governance and anti-incinerator campaigns in Ireland, North and South

Liam Leonard[a], Peter Doran[b] and Honor Fagan[c]

[a]School of Business & Humanities, Institute of Technology, Sligo, Ireland; [b]School of Politics, Queens University, Belfast, Northern Ireland; [c]Department of Sociology, National University of Ireland, Magnooth, Ireland

The decades of conflict in Northern Ireland created divisions between communities, with few opportunities for cooperation. However, in the 1990s opposition to a proposed cross-border incinerator brought the divided communities together. The 1990s economic boom in the Republic of Ireland generated a waste management crisis as the by-products of rampant consumerism overwhelmed the state's rudimentary waste disposal system. Three Irish anti-incinerator campaigns which have pitted local communities against the Irish state or the Northern Ireland Department of the Environment are examined. Community attempts to gain leverage within the political governance frameworks in operation on both sides of the border are examined and the various ways in which environmental movements respond to the crisis of waste management under different governance regimes are illuminated.

Introduction

Anti-incineration campaigns emerged on the island of Ireland during a particular phase of economic growth with increased consumption in the Republic alongside conflict resolution in the North from the late 1980s through to the turn of century. The political frameworks that existed on both sides of the border whilst these changes occurred are based on partnership agreements between national and regional political entities or the local authorities and the corporate sector. This wider political opportunity structure included the European Commission (EC), and in the Northern jurisdiction, the UK government, the Northern Ireland Office and the Department of Environment

(DoE), as well as the post-1998 devolved Stormont Assembly and, in the south, the government in Dublin and the waste management industry. Those opposed to incineration technologies included a loose coalition of citizens with concerns about health risks, environmentalists and, in the Northern case, both nationalists and unionists who had historically been trenchant opponents on almost every other local issue.[1] We will argue that the partnership arrangements that existed in both jurisdictions, but to different extents, from the outset excluded communities that opposed incineration from the core of political decision-making. Their campaigns achieved various degrees of political access with differing consequences. We examine the dynamics and features of three campaigns against incineration in the two political jurisdictions in their governance context.

Following Rootes we note that community-based environmental movements ebb and flow within local frameworks:

> local environmental campaigns are ubiquitous and recurrent, even in times when environmental issues are not salient on national agendas ... Local environmental campaigners are variously related to national and local organisations, and the peculiarities of place are one factor in that variation. But place itself acquires meaning through campaigns, and communities forge identity even as they mobilise against threats to their survival. (Rootes 2007, p. 722)

Our focus here is on local environmental movement activism in the context of waste infrastructure and its political governance. We will examine the movement dynamics of three regional campaigns that occurred in the 1990s and early 2000s in Derry, Galway and Cork regions. A waste management crisis emerged as a result of increased consumption rates as economic growth was achieved in the Republic and as a post-conflict process was emerging in the North. European directives led to a shift from a traditional (over-)reliance on landfill, with regional waste management plans including incineration as an option for the first time. We also outline a series of dynamics in the mobilisation of networks for the anti-incinerator and zero waste movements, the role of experts and advocates with international experience of similar campaigns, the significance of cultural capital in the mobilisation process, and the differences in rural and urban campaigns. Finally we examine the extent to which political access was achieved by community groups originally alienated from the diverse systems of political partnerships in both jurisdictions.

Governing waste

Waste management issues and the broader issue of sustainable development are, ultimately, issues concerning *governance* in contemporary societies. Governance is taken to refer to the sum of interactions between civil society and governments. Good governance is currently taken to include transparency, effectiveness, openness, responsiveness and accountability. These are all criteria by which we can legitimately evaluate the government of the day in respect of its dealing with civil society and its concerns. So, if sectors of civil society –

concerned individuals, community groups and environmental groups – have concerns about the government's waste management strategy (whether as discrete projects or as an indicator of the broader concern with sustainable development in particular), we can expect them to be dealt with according to the internationally accepted criteria of good governance.

The essence of governance is its focus on mechanisms to govern society which do not rest on the use of authority or of sanctions by government. Forms of partnership with the organisations of civil society are one of the preferred options of the governance approach. Following one of the authorities in this area, 'governance recognises the blurring of boundaries and responsibilities for tackling social and economic issues' (Stoker 1998, p. 21). Both in terms of strategic decision-making and of service delivery there is now a growing critique of the 'Westminster model' (see Bache and Flinders 2005, Bevir 2008) on the grounds that its centralised powers lack flexibility and the necessary counterbalances. Government by central decree is becoming increasingly unpopular in contemporary western society, and is often replaced by a more consensual model based on multi-agency partnerships or some hybrid model. The notion of governance (and particularly multi-level governance in an EU context) responds to these perceived weaknesses with an ideal type stressing the complexity of modern-day political management and the need for ongoing citizen participation in that process.

When we look at the production and management of waste it is useful to think in terms of multi-scalar processes, where rescaling of waste production in the era of glocalisation has occurred and where its successful management relies on governance at multiple levels – global, regional, national and local. EU directives on waste have been the key driver of waste management policy in Ireland (Fagan *et al.* 2001). The European Economic Community (EEC) Act of 1972 gave direct precedence to European acts over domestic laws and constitutional provisions in the Republic and in Northern Ireland. The ratification of the Single European Act (1986), the Treaty of Maastricht (1992) and the Treaty of Amsterdam (1997) further ensured the supremacy of EU law over domestic law. In its programme for dealing with waste the EU produced legislation, which includes Directives on dangerous substances, waste oils, groundwater, urban waste water, licensing regulations, the disposal of PCB/PCT, toxic waste, sewage sludge in agriculture, emissions from waste incineration plants, the disposal of animal waste, and batteries containing dangerous fluids. It likewise set targets for reduction in all waste streams, and set very specific timeframes for national governments to meet these reductions. For example, for the Republic of Ireland's municipal waste stream, there is a national target of 35% recycling by 2013 and a household waste diversion from landfill target of 50% by 2013.[2]

The state governance context

With the EU able to enforce sanctions on the nation-state and the national government needing to radically change the direction and composition of waste flows, the drawing up and implementation of strategy quickly became an issue of governance at a national level. Government by central decree on the waste management issue was not an option since the government had moved to a governance model patterned on consensual politics and multi-agency partnerships. From this perspective, self-governing networks in relation to waste management were very much favoured by the state. The capacity to 'get things done' did not simply rest on the power of government to command, and commands would only be invoked in the last instance. In order to reach the targets it was considered necessary to bring key players such as 'private enterprise' into some form of partnership. In 2001 in the South there was a need for an estimated investment of €1 billion over a 3–5 year period to implement the waste development plan (Forfás 2001, p. vi) and the National Development Plan envisaged this coming mainly from the private sector.

Clearly, Ireland faced a gruelling task to organise a strategy to divert waste away from landfill, to reach targets set at a five-fold increase in recycling and to find the finance for the infrastructure, especially if the objective was for the private sector to answer this call. Private capital was thus seen a necessary 'node' in the governance of waste management (Fagan 2004). Offe's (1987) argument that the neo-corporatist system focuses on 'technocratic criteria' is illustrated by the policies generated under these 'crisis' conditions. In such conditions, the government's gaze focused on the private sector and on the waste industry's multinational giants, leaving sustainability concerns secondary to costs and slow, deliberatively reached democratic solutions secondary to immediate technocratic solutions. Waste governance in Ireland, from this perspective, could not be resolved at its most radical level – that of sustainability. The plans relied heavily on the treatment of waste through 'thermal treatment plants' and on recycling to be funded primarily by private enterprise. In short the partnership arrangements involved a hierarchisation of partners, with the role of the community or civil society partners being to consent to the technocratic logic of 'getting things done'. Partners offering solutions to the crisis and engaging in problem solving were 'more equal' in the partnership than those who could be labelled 'oppositionists'. Clearly the way governance is implemented is part of this political process and should not be seen as a *deus ex machina*. Being oppositionist brings with it a danger of exclusion from partnership arrangements. Murray (2006) argues that such multi-level partnership arrangements involving state and corporate entities may serve to further isolate working class communities and also marginalise them from the local community perspective. It is perfectly rational that in a 'problem-solving partnership', 'partners' offering solutions to the crisis and engaging in problem solving would be 'more equal' in the partnership than those labelled 'oppositionists' and risking exclusion from the prevailing partnership arrangement (Gaynor 2008).

While governance necessitated consultation and the introduction of key players into the process, the unequal balance of power in the consultations and the fact that some partners were 'more equal than others' resulted in outright contestation of the plans in the South, where plans were about a year in advance of those in the North. How did communities and activists contest their respective government's preferred waste management strategies? Discursively they contested the strategy as 'bad' governance which they located in what they termed a 'non-consultative' process. The environmentalists and local communities threatened by incineration plans were deeply critical of what they perceived as the 'façade' of consultation that had been put in place (Fagan *et al.* 2001, p. 18). There was a widespread perception at community level that government 'consultations' (often dictated by EU regulations) on the development of incinerators were simply empty rhetorical exercises for communities to 'let off steam' but were not designed to change decisions already taken on technical grounds (Fagan *et al.* 2001, p. 19).

In the North, the proposed all-Ireland incineration plant for toxic waste was blocked by cross-border and cross-community opposition. In the South the opposition to the location of incineration plants began, fuelled by anger about the nature of the consultation process that had produced the plans, and drove the waste management strategy into political crisis in 2000–2001 as local communities blocked the sub-regional plans. In the North, the failure of the proposed DuPont plans for an all-Ireland toxic waste incineration plant drove cross-border cooperation into the background. A second development that minimised cross-border cooperation was, curiously, the impact of the political backlash to the waste management plans in the South. Alongside the opposition to the DuPont plant, that opposition to plans for large incinerators in the South created a situation where the Northern planners watched with horror as they saw the backlash against plans for large-scale incineration plants unfold. Not wishing to go down the same road, waste authorities in the North proposed many small incinerators instead and community buy-in was considered to be a priority. Given, therefore, that there appeared to be no urgent reason for anything but small-scale regional planning, cross-border cooperation went on the backburner and no one seemed to be actually working on developing an all-island strategy.

The Southern state moved into action against the blocking of the plans by anti-incinerator activism. The first Environment Minister to deal with the issue, Noel Dempsey, removed local councillors (who had been subject to public will) from the decision-making process, and replaced them with the county manager, a government employee. So, in response to challenge from 'below', a central decree (government as opposed to governance) was used to achieve the localising or embedding of waste management. This is not to say that the Irish state entirely moved back to traditional government or rejected the principle of consensus politics and failed to involve itself in multi-agency partnership, but, rather, that it removed the locality from involvement in the decision-making process. The next Environment Minister, Martin Cullen, stated quite openly that the planning process on waste management was 'over-democratised' and that he did not believe it was 'adding anything to it by having so many layers involved' (*Irish Times*, 12 August 2002, p. 1).

The so-called 'fast-tracking' for waste management plans had to be implemented, and An Board Pleánala (the Planning Board) became a 'one-stop shop' for assessing all plans for new waste management facilities. The Minister, rather contradictorily, insisted that he was not removing the rights of any groups or individual to express their views – 'That is sacrosanct, but I don't see a need for these views to be expressed at so many different levels' (*Irish Times*, 12 August 2002, p. 1). In other words, repetition of oppositional views at multiple levels in a multi-layered process of governance was a source of irritation for government.

This suggests a particular multi-faceted and shifting dynamic of actors in the governance process, with some gaining power and others losing it in a complex political process of action and reaction. The discourse of governance certainly did not ensure that the political will of communities would prevail. That local communities were important players in the dynamic is without question, but there were ebbs and flows in their political power. Let us look in more detail at these in the three case studies that follow. In terms of governance, the EU is a key player in that it regulates waste and sets the scene for its regulation at national level. However, EU policy emerges from a network of actors and competing agendas and is translated into national policy through a similar network. While we can clearly see the European agenda informed by concerns with environmental sustainability, we can equally see the market-driven notions of development being played out when it comes to its implementation at national level. At the implementation stage the contradiction between the concepts of economic development (market-driven in its capitalist form) and sustainability (the earth as limited resource) play themselves out, with the former being presented as more urgent than the latter.

Ireland: the North–South governance context

Prior to the Belfast Agreement which brought an end to the conflict in Northern Ireland, cross-border cooperation had existed, as we will see in the case of the DuPont proposal to site an all-Ireland toxic waste incinerator in Northern Ireland. The DuPont campaign took place at a critical period in waste management politics, North and South. Fagan *et al.* (2001) noted that in both parts of Ireland waste management was entering a critical moment in light of new EU regulations and the attitude of many local communities to incineration in particular. European Directives that required governance consistent with enhanced environmental sustainability were affecting industry, commerce, local authorities and households in an increasingly direct way. After intensive negotiations between the Irish and British governments (and the various political parties), the specific nature and administrative form of North–South cooperation was put in place in December 1998. Six North–South implementation bodies were established, covering waterways, food safety, trade and business, EU programmes, the Irish language and agriculture/marine matters. Also, the so-called Trimble–Mallon statement in December 1998 contained an initial list of six matters for North–South cooperation through existing North–South public policy bodies. These were to include transport, agriculture, education, health, tourism and the environment, the latter specified to include research into environmental protection, water quality and waste management. So waste management in Ireland became part of an all-island public policy framework. But we can also note that the Du Pont case pre-political settlement had already refigured this outcome as waste management was already being seen as an issue transcending political borders in a small island.

Policy-makers on both sides of the border were facing the introduction of significant targets, including a considerable reduction in the amount of waste going to landfill, and increased energy recovery from waste. In 1988, the Department of the Environment in the South commissioned consultants to prepare a feasibility study on a 'national' toxic waste incinerator. By 1990, consultants on both sides of the border had advised their respective governments of the desirability of an all-island solution to the challenge of disposing of toxic waste. Key to considerations on both sides of the border were developments at the European level. The ratification of the Single European Act (1986), the Treaty of Maastricht (1992) and the EC Waste Strategy launched in 1989, enshrining the principle of sustainable development, combined to force governments in Dublin and Belfast to make waste management a domestic priority. The principle of 'self sufficiency' in dealing with waste disposal was established in European legislation in 1991 and economies of scale pointed to the need to cooperate. As a Northern Irish civil servant put it succinctly in 2001 'the whole North/South thing is where [waste management] is going to happen', and according to him two things were fundamental: 'one is scale and the other one is working together' (Fagan *et al.* 2001, p. 44).

Case study I: DuPont, Derry (Northern Ireland)

In one of the island's most celebrated campaigns against a proposed toxic waste incinerator, communities in Derry and Inishowen took on Derry's biggest and most important employer, DuPont, at Maydown on the outskirts of the city. During the 18-month campaign (1990–1991) a question of environmental justice merged with the dominant concerns of citizens and communities emerging from a deep political conflict and history of communal division. Derry – sometimes regarded as the crucible of the civil conflict and nationalist/ Republican assertion of their civil and political rights – provided the *mise en scene* for a successful campaign, which would tap into a number of discourses drawing from a distinctive political culture and set of historical circumstances. Not least of these was the ease with which Protestant and Catholic, Unionist and Nationalist, citizens and politicians contradicted the dominant narrative of division and conflict to make common cause in their assertion of the city's and region's right to say 'no' to what was widely regarded as an imposition cooked up by outsiders, including corporate actors at DuPont and government ministers in both Dublin and Belfast. This case study provides valuable material for an understanding of how divided communities might mobilise around environmental issues and the way new environmental discourses are constructed on pre-existing community consciousness. The name of the DuPont company was already known on both sides of the border when the Derry campaign was launched. In 1974, the company had given consideration to the establishment of a plant at Cork Harbour to extract titanium dioxide from limonite ore from Australia. Controversially, the company had proposed to dump the resulting waste in the open sea. Sixteen years later, the anti-incinerator campaign took place at a critical period in waste management politics, North and South.

The origins of the proposal for an all-island incinerator were recommendations set out in a feasibility study commissioned by the Republic's Department of the Environment. In turn, consultants commissioned by the Department of the Environment in Northern Ireland recommended that the Department engage with its counterpart in Dublin. This was the background to a visit by Flynn to the DuPont plant in 1990. The company engaged the Irish Government with a view to importing 14,000 tonnes of the proposed incinerator's projected 20,000 tonnes annual capacity for treatment at its Derry facility. The company ruled out 'importing' waste from Britain to the largely nationalist city of Derry. Allen (1992) concluded that the consultants' suggestion that there should be a cross-border solution to the toxic waste challenge which had bedevilled successive Dublin governments seemed to offer a way out for the Irish state.

Reviews of waste strategy on both sides of the border had coincided with DuPont's plans to upgrade its incineration capacity at the Derry plant. Economic considerations for the industry, including the prospect of grant aid to DuPont, merged with the logics of the European Commission's first broad-based communication on waste, *A Community Strategy for Waste Management* (1989), including the 'proximity principle'. The consultants advising the Department of the Environment in Northern Ireland, Aspinwall and Co. Ltd, for example, noted that the waste markets on the island could not support incinerators in both Northern Ireland and the Republic, unless operators were to consider importing wastes from other countries. The consultants noted that proposals for a single viable incinerator servicing the waste markets in both jurisdictions would fall within the spirit of the 1989 Waste Strategy, especially since DuPont had each year been exporting 700 tonnes of hazardous waste to Finland and France for incineration.

Activists from Derry and Donegal reacted with shock when they learned of the proposals to site an all-island incinerator at the DuPont plant, five miles outside Derry, in early 1991. Local political representatives were also quick to respond to the announcement and an emergency meeting of Derry City Council's Environmental Protection Committee convened. The Social Democratic and Labour Party (SDLP) group, which controlled the Council at the time, relayed the 'alarm and concern' among local politicians. While consensus emerged amongst Nationalist and Unionist politicians, their opposition was based on a variety of arguments. Some Unionist politicians, reflecting the deep resentment stirred by the Anglo-Irish Agreement, opposed Dublin's interference. Sinn Fein representatives, with close ties to grassroots community-based organisations, were quick to articulate technical, economic and social arguments against incineration. The grassroots campaign, for the most part, maintained a firewall between cross-community and cross-border activism and the role of local politicians, and elected representatives from political parties were not encouraged to address public rallies.

DuPont's proposal was to take toxic waste from across the island to ensure the technical and financial viability of the new plant. The imported element would come from chemical and pharmaceutical companies based in Dublin and Cork. Du Pont launched Northern Ireland's biggest ever industry-sponsored public information campaigns to explain its plans and offer reassurance to the local population. The Derry population was not known for its interest in environmental issues (Allen 1992, p. 3). Key drivers of the initial campaign formation were prominent activists based in Donegal, just across the border, in the Republic of Ireland. A feature of the cross-border campaign was the distinctive mobilising narratives deployed on each side of the border. While established community-based environmental lobby groups in Inishowen, Donegal, focused on the perceived pollution threat, the debate in Derry focused as much on issues of the rights of the community to participate in decision-making, failures in environmental governance, and the economy. The Derry-based campaign thus resonated with that city's historical identification with public protest in pursuit of civil and economic rights.

The cross-border campaign was launched at a meeting in Derry in January 1991. Within 24 hours, at a meeting of Derry City Council's Environmental Protection Committee, opposition to the incinerator proposal was made clear. A campaign 'Briefing Paper' (1991) drawn up by the Derry Development Education Centre set out a number of influences on the campaign and key arguments that place the events in context, on the eve of the UN Conference on Environment and Development. The major considerations for campaigners included: the 1972 London Convention to outlaw the dumping of waste and incineration in the North Sea; a local history of industrial toxic waste disposal on farmland; a perception that accommodating industry's preference for incineration could reduce pressure to address waste prevention; the risk of large amounts of toxic waste being transported to Derry and the risks associated with Derry becoming a magnet for all future toxic waste generated on the island; the scientific uncertainties surrounding the 'cause–effect' relationship between industrial pollution and public health; a perception that pollution monitoring had been 'piecemeal'; and a failure to enforce emissions regulations and punish those in breach of regulations.

Having noted that the proposed site for the incinerator was closer to villages in the Republic of Ireland than to the city of Derry, campaigners discussed their suspicions about atmospheric pollution drifting across Lough Foyle from Derry to Donegal. The Inishowen Environmental Group in Donegal tapped into these suspicions to mobilise public support for the anti-incinerator campaign, claiming that marine and air pollution from industrial activity in Derry had already resulted on both sides of Lough Foyle.

The focus for the mobilisation of groups in Derry and its environs was a development education centre, with trade union ties and funding. The Derry Development Education Centre (DDEC) played a key role in providing a secretariat for a network of 63 local groups, many of them specifically set up in neighbourhoods for the purpose of opposition across the city and on both sides of the border. The Centre exploited links with the trade union movement, international development and environment NGOs, and industrial health and safety networks. The DDEC worked alongside veteran community develop-ment organisers, some with a history of campaigning and mobilisation stretching back to the early civil rights movement in the city in the late 1960s. Conventional campaign tactics were deployed, including public rallies, a media strategy, lobbying, theatre and fund raising.

The campaign counted support from local bishops, elected representatives, the media, farmers' groups, including the Ulster Farmers Union, community groups, doctors, trade unions and women's groups. They drew on advice from a number of professionals within their ranks, including energy consultants, engineers, marine biologists, farming experts, architects and health workers. The campaign also had significant access to research and expertise from across Ireland, the United Kingdom and the United States. Greenpeace advisors offered support on the media strategy and tactics.

Timeline for the Northern campaign

November 1988: Then Republic of Ireland Environment Minister, Padraig Flynn, commits to incineration as a long-term solution to the chemical industry's waste problem, speaking at a conference in Cork. He commissions consultants, Byrne O'Cleirigh, to prepare a feasibility study on a 'national' toxic waste incinerator.

February 1989: Report by the consultants, Byrne O Cleirigh, commissioned by Flynn. Tenders for a national incineration project invited in the Republic of Ireland.

June 1990: Consultants Aspinwall and Co. Ltd present findings on the future of waste disposal in Northern Ireland, in a report commissioned by DoE NI. They include a recommendation that NI Ministers undertake joint discussions with their counterparts in the Republic of Ireland. The talks would determine the feasibility of including Northern Ireland in the Republic of Ireland's proposals for high temperature incineration. The consultants' report cited the EC Waste Strategy (1989).

December 1990: With no decisions taken on tenders for the Republic of Ireland facility, Flynn visits Du Pont in Derry, to pursue possible cooperative arrangements involving the administrations in Belfast and Dublin.

5 December 1990: Du Pont submits application for planning permission to replace a solid waste burner to burn lycra waste. (Du Pont never submitted a request for permission to develop the proposed all-island facility.)

18 January 1991: The Inishowen Environmental Group in Donegal poses key questions about DuPont proposals in an article published in the *Derry Journal* newspaper.

21 January 1991: 120 people attend an anti-incineration briefing and launch of campaign in Derry's Central Library.

21 January 1991: Greenpeace/Cork Environmental Alliance meets Derry City Council officials. At least one solidarity rally was staged in Cork to coincide with a protest convened in Derry during the 1991 campaign.

30 January 1991: Formation of Campsie Residents against Toxic Emissions, involving residents living in the environs around the Du Pont plant.

31 January 1991: The Economy Minister for Northern Ireland, Richard Needham MP and the Environment Minister for the Republic of Ireland, Padraig Flynn, meet at an Anglo-Irish conference in Dublin and discuss the proposal for an all-island facility at DuPont.

16 March 1991: First anti-incineration rally in Derry.

20 August 1991: Du Pont announces that it is to begin an environmental impact study (EIS) in preparation for a planning application for an all-island toxic waste incinerator.

December 1991: DuPont announces that it is abandoning plans to build an all-island incinerator at its Maydown plant in Derry. The decision, which had been taken out of the hands of the Derry plant operators, is attributed to strategic financial considerations.

Ultimately, DuPont announced in December 1991 that it was abandoning the incinerator plan. Underlining the importance of local autonomy and democratic decision-making as a feature of the campaign, the company attempted to signal that the final corporate decision had been taken at an international level 'so that both the responsibility for and the cause of deciding not to proceed appeared to be taken out of the local arena' (Jordan and Gilbert 1999, p. 2).

An editorial published in the main local newspaper, *The Derry Journal* (3 May 1991), had come out in strong support of the anti-incineration campaign and neatly located the campaign's arguments within the historical and political context of the city. The editorial stated that there were good reasons for the opposition given that the city had suffered enough from discrimination from Stormont governments over two generations. Local people, it continued, were in no mood to allow the installation of a plant they believed would damage its environmental attractiveness and stunt hopes for an economic revival. The editorial concluded with a warning to both the Northern Ireland Office and the Dublin Government: 'The dismissal of community protests from Derry played a major part in precipitating the civil rights movement in the North. The toxic incinerator has become the biggest issue to command support across political divides since those times. Both the Northern Ireland Office and Dublin should tread warily.'

Frames and meanings

Jordan and Gilbert (1999) have studied the 'competing discourses' or 'distinctive social meanings communicated through the use of language' which were at work during the Derry incinerator dispute. They concluded that while both supporters and objectors shared the language and discourse of 'environmental management', only the objectors mobilised around 'environ- mental representation' and this imbalance contributed significantly to the defeat of the proposed incinerator. As Jordan and Gilbert (1999) noted, unlike expert discourse, the discourse of 'environmental representation' does not rely upon a strict bifurcation of 'environment' and 'community'. The objectors successfully established legitimate claims by drawing on this discourse, rhetorically tying a particular social grouping or community to a particular physical space and environment (Jordan and Gilbert 1999, p. 5), as captured in one quotation recorded in the *Derry Journal* (27 September 1991):

> Too often the word 'environment' is used in a narrow and restricted sense as almost exclusively confined to the physical surroundings ... If the environment includes all aspects of surroundings then the cultural and social dimension is just as important as the physical and the geographical and must be included in any environmental impact assessment.

Emerging from conflict, and anticipating a 'peace building' era of new choices about infrastructure, the economy and the nature of local development, the DuPont campaign forced open debates that other parts of the island had already rehearsed. Environmental and health considerations were framed by a strong sense of local community. For historical reasons, activist com- munities in both Derry and Inishowen were wary of remote government decision-making and a perceived neglect in terms of economic and develop- ment outcomes. This gave a very local and context-specific flavour to argu- ments around access to information and participation in environmental decision-making. The political environment created by the highly contested Anglo-Irish Agreement deepened suspicion on the Unionist side, while on the nationalist side promises of job creation came to be seen as an attempt by the authorities to exploit Derry's status as an unemployment blackspot rather than a *bona fide* attempt to redress an historic injustice.

Jordan and Gilbert (1999, p. 10) suggest that the Derry dispute was essentially a political debate concerning human relations. Protagonists used ostensibly 'environmental' concepts to frame arguments concerning appro- priate social relations, in this case about decision-making over local develop- ment. One letter writer to the *Derry Journal* (22 January 1991) was quite explicit in challenging the paper's editorial writer to accept that a 'community values' perspective was just as valid as that of scientific expertise, which must be assessed given that 'the history of science and technology, or rather the history of the use of science and technology reveals an absence of objectivity and very strong value loading'.

Case study II: Galway for a Safe Environment (Republic of Ireland)

The regional waste plans for the Republic of Ireland included three options: landfill (which was the destination for over 90% of the country's waste), recycling and incineration. The inclusion of plans for an incinerator at certain named locations in and around the western city of Galway caused local professionals to instigate a campaign of opposition to the siting of the plant, while their campaign would emerge into a wider anti-incineration campaign with extended links regionally and globally, while attempting to influence the 2002 General Election (Leonard 2005). The local campaign of Galway for a Safe Environment (GSE) opened up three main frames: highlighting health risks; embracing wider politics and going 'beyond NIMBY'; highlighting democratic deficit.

These three frames were not distinct, and tended to overlap as GSE's leadership attempted to politicise their campaign by moving 'beyond NIMBY' (Szasz 1994). The initial phase of the campaign gave rise to a series of protests, marches and media appearances that allowed GSE to highlight the issue of health risks posed by incinerator emissions. GSE's health frame provided many potent images for the anti-incineration activists to manipulate in order to create issue salience amongst the public. All aspects of community politics were integrated into GSE's anti-incinerator repertoire, including exploiting anti-abortion sentiment still prevalent in the wake of two referendums on that contentious issue. (GSE promoted the image of dioxins in baby's milk as one of the main health risk concerns.) Furthermore, GSE outlined the damage caused to processes when pastures and farms were exposed to toxins including furans and dioxins in emissions from incinerators. This was done in order to exploit another cultural frame based on existing concerns about toxic pollution from multinationals in rural areas.

In so doing, GSE was able to extend the cultural frame to embrace rural environmental sentiment, while also preventing a rural/urban divide, something that would have benefited its opponents (Leonard 2008b). This strategy emerged from the prior experience of some GSE committee members, who had knowledge of anti-incinerator campaigns in Canada. These links to international anti-toxics campaigners such as Professor Paul Connett of St. Laurence University, New York, would provide GSE with a vast resource of scientific data that provided the basis of their health frame. In fact, GSE were able to provide a great deal of information on incineration to the public, local politicians and media sources, to the extent that the interest-driven data came to shape the debate, with the state and industry being forced into a reactive stance. The forms of action taken by the Galway campaign included a mix of street protests outside of City Hall, with spokespersons addressing alternatives such as the recycling based 'Zero-Waste' approach to waste at special meetings of the local authority. In this way, the Galway campaigners were able to maintain pressure on the local political structure, while providing a level of expertise to the local (and national) debate through their wider networks. At the height of its campaign, GSE was holding major public meetings debating the issue live on the evening news while its petition against incineration received 22,000 signatures in a city of 70,000.

This mobilisation of support was also reflected in the extent to which GSE influenced local councillors, who went on to reject the regional waste plan. Many councillors publicly stated that GSE's campaign had influenced their decision, while many reported an upsurge in voter concern on the issue. The state's response to this rejection of its waste policy was to rescind the decision-making powers of all regional councils on waste management issues, a move that provided GSE with the political opportunity of extending the democratic deficit frame. For GSE and its supporters, the state's initial approach to pushing through incineration without consultation or referring to any potential health risks in the regional plan was one example of a lack of accountability on transparency on the issue. However, the removal of the councillors' decision-making powers allowed GSE to re-frame the campaign by attempting to gain wider access to the political structures on a national level. The key opportunity for that strategy presented itself through the 2002 General Election. As the balance of political opportunities (another significant aspect in determining campaign outcomes) surrounding the anti-incinerator campaign continued to shift, GSE was able to extend the democratic deficit frame, gaining further leverage during the general election campaign. Having decided against running its own candidate in order to facilitate supportive political figures from the mainstream, GSE began to merge its three main frames into an offensive against Fianna Fáil, the main party of government. Moreover, while Fianna Fáil had targeted three seats in Galway West, GSE created strategic alliances with one government party (Progressive Democrat – PD) candidate and one opposition party (Greens) candidate to increase the chances of having a supporter in government (Leonard 2005). Ultimately, the return of the Fianna Fáil/PD coalition to power dealt a major blow to GSE's attempt to politicise its campaign. Essentially, GSE's key political alliances had proved to be no more than a 'perceived' opportunity (Tarrow 1998), rather than the key leverage that would lead to the campaign influencing policy at a national level. Nonetheless, while municipal incinerators for Cork and Meath (near Dublin) were announced in November 2005, any such plans for Galway had been shelved, with Fianna Fáil keeping one eye on the potential for populist backlash in future elections.

Case study III: Cork Harbour for a Safe Environment (Republic of Ireland)

Another site in the Republic, Cork Harbour, was part of a series of municipal incinerators announced by the state as part of seven regional waste plans in Dublin, the South-East, Galway, Limerick, the Midlands and the North-East

(Fagan 2003, p. 67). Communities in these areas campaigned against the proposed incinerators, using collective action frames to create understandings about the potential health risks posed by incineration. The Cork campaign had marked similarities with GSE's, as political opportunities were provided by the state's technocratic approach to including mass-burn incineration:

> One can see that incineration is the contested terrain in this case, as not one government policy or regional plan mentions the word 'incineration.' The word used repeatedly and pointedly is 'thermal treatment plant.' As in all conflicts, the discourse itself marks the terrain and the use of the work 'incineration' as opposed to 'thermal treatment plants' marks the political division. (Fagan 2003, p. 78)

Moreover, the state's own attempt to frame the waste issue in a less than up-front manner provided the first political opportunity for community campaigners challenging the regional waste plans. In December 2001, Indaver, the Belgian incineration company involved in all of the regional plants, lodged a planning application for a 100,000 tonnes per annum incinerator for hazardous and non-hazardous industrial waste, as well as an incinerator of equal size for municipal waste, at a site at Cork Harbour. This incinerator was opposed by activists in the Cork Harbour/Ringaskiddy area, which had been the site of previous disputes about toxic dumps and industries. This provided their campaign with further leverage due to the existence of many activists who had taken part in previous campaigns. As such, the resources of campaign and scientific expertise could be more easily mobilised. The proposed incinerator was opposed by local residents, who formed an alliance from the many existing environmental groups in the Cork area, under the name of the Cork Harbour Alliance for a Safe Environment (CHASE). The campaign gained momentum, and CHASE was able to claim to 'represent the views of 24,000 people' (*Irish Examiner*, 19 September 2003). CHASE followed similar strategies to Galway for a Safe Environment, by establishing a network of environmental groups, mounting a campaign of public protests and meetings. However, CHASE had one major difference in its strategic approach: it focused on a direct challenge to Indaver, rather than attempting to influence the political process. Indaver and the state were keen to site their incinerator at Ringaskiddy, Cork Harbour, as '60% of all hazardous waste in Ireland comes from Little Island and Ringaskiddy' (ibid.). Links were established between CHASE, GSE and the No Incinerator Alliance in rural County Meath, as resources and expertise were exchanged between the groups.

The CHASE campaign saw the appearance of many of the women and children in white boiler suits and oxygen masks, holding 'No to Incineration' placards, and presenting a striking visual profile (*Irish Times*, 17 January 2004). CHASE was able to frame the 'moral discourse' (Grove-White 1993, p. 20) surrounding the issue, depicting the children in a way which highlighted their potential exposure to the health risks of dioxins. The dispute was taken to a series of oral hearings, held at the Neptune Stadium in Cork. In this way, the campaign led by CHASE manifested itself in a manner that had become the norm for environmental disputes in Ireland. As legal challenges may only provide 'partial access' (Tarrow 1994, p. 86) but entail high costs, oppositional campaigners have less often had recourse to law than the EPA and the government. This is the reason such a strategy was avoided by Galway for a Safe Environment. Such strategic decisions provided GSE with a level of success, in contrast with CHASE. After a campaign lasting over two years, An Bord Pleanála (the Planning Commission) refused permission for the Ringaskiddy waste scheme on the grounds that the Environmental Impact Statement was inadequate, an incinerator for non-hazardous industrial waste was contrary to the Cork Waste Management Plan, would contravene materially the zoning of the site primarily for industry and enterprise, the scale, nature and purpose of the development would be fundamentally unsuitable to the site and 'that it would be close to high density housing'. These were amongst 14 points given for the refusal of planning permission at the site (*The Irish Times*, 17 January 2004).

The senior planning inspector's report ran to over 300 pages, and rejected Indaver's plan for an incinerator on the 14 counts. CHASE claimed a complete vindication of its position on the issue, as many of its objections formed part of the inspector's rejection of the incineration scheme. However, both the inspector and CHASE were soon to find that An Bord Pleanála would overrule the refusal and grant planning permission for Indaver's proposed plant, citing the prioritisation of government waste policy over all other considerations. An Bord Pleanála also had to consider the state's waste management frame-work, in which there is a preference for incineration over landfill. The board also considered the geographical location of a number of large-scale chemical and pharmaceutical industries operating in the Cork Harbour Area, and stated that Indaver's plant was 'an appropriate location for a necessary public utility' (*Irish Times* 17 January 2004). It also stated that such a plant would 'not injure the amenities of the area or be prejudicial to the future for port-related development' (ibid.). The then Green party representative for the Cork South Central constituency, Dan Boyle, said this decision was a 'political' one, and accused the board of 'caving in' to government pressure. He questioned the fairness of the planning process, adding that the campaign against the plant would continue: 'This battle now moves to its next stage, where many thousands who have campaigned against this incinerator will look for something in the planning, political and judicial process that can deliver them true justice' (ibid.).

However, the state's response to CHASE, which included the granting of a limited degree of access through the oral hearing, followed by the overriding of An Bord Pleanála Inspector's own findings, represented a reversal of fortunes for the CHASE campaign, and indicated the limitations of the oral hearing approach for public campaigns, something which has been noted about previous community campaigns in the Cork region (Peace 1997). The CHASE campaign demonstrated how creating 'increased access' (ibid.) and the ability to create 'alignments' with existing environmental activists in the Cork area, together with limited support from allies such as An Bord Pleanála's inspector, provided assistance. Nevertheless, 'the strength of state' (Tarrow 1994, 1998) was the decisive factor, as the state overruled the Board's inspector in the interest of pursuing its wider policy aims.

Conclusion: growth, waste and peace in the new Ireland

We have examined the mobilisation and constellations of political opportunities surrounding the governance of waste in both jurisdictions in Ireland. Each of the campaigns met with differing outcomes, due to the nature of the political frameworks in those jurisdictions. The three Irish case studies presented here demonstrate differing responses by social mobilisations to the state's governance of waste management. Despite what many consider to be the two political success stories of recent times in the two jurisdictions – the 'peace process' and the establishment of a devolved, consocial Assembly in Northern Ireland, and the rise of the neo-corporatist 'social partnership' arrangements which underpinned the 'Celtic Tiger' economy in the Republic – local communities have suffered severe fears about the risks of incineration and are unable to have their concerns dealt with without resorting to social mobilisation. While the major political agreements represent a considerable achievement, they have created a degree of exclusion for those groups that remain outside the socio-economic and political elites that have emerged in Belfast and Dublin in the last two decades.

Davies (2007) and Garavan (2007) have examined the impact of environmentally orientated community movements in Ireland. Despite having local political access and expertise, community campaigns have been excluded from the political core in both jurisdictions due to the influence of neo-corporate partnerships that prioritised growth over community concerns in the South, and the nature of the consocial but politically weighted Assembly in the North, which was developed to accommodate the sectarian political divide rather than environmental or other 'social' concerns. In the main, these groups are representative of those without power under the new political dispensations. Non-economic actors such as women's groups, the unemployed, immigrants and environmentalists have been marginalised by the coalition of mainstream parties, trade unions, large farmers' groups and the business lobby in the Republic, and in the North by most of the political parties, whose focus is on controlling the levers of power in the Assembly.

In the Republic of Ireland, there is a tendency within the existing neo-corporatist system to focus on the technocratic aspect of policy-making capacity which creates difficulties for political parties who wish to represent the concerns of the professional middle class, due to the tendency of professionals to have concerns that go beyond the economic. This void can be filled by the campaigns of environmental movements, as public sector professionals such as academics and those with alternative forms of expertise mobilise and challenge the structures of closed corporatist power. The critique of closed political systems in European states such as Germany and Austria can be applied to Ireland's own system of neo-corporatist closure with the question: 'What is it about neo-corporatist arrangements that have stimulated the development of Green movements in those countries' (Scott 1990, p. 144)? While this question may not be as relevant in the Irish case, the persistence of environmental campaigns and the emergence since 2007 of the Irish Green Party as partners in a coalition government dealing with the incineration issue provides some congruence with European Green movements.

Corporatist and neo-corporatist centralised arrangements and the processes of inclusion or exclusion that result from the state's facilitation or repression of access to political structures creates a 'dimension of political opportunity' (Tarrow 1994, 1998) for movements such as environmental campaigners. 'Inclusive' corporatist arrangements are usually 'restricted to employers and organised labour' (Scott 1990, p. 144). The restricted nature of corporatist arrangements 'means that groups excluded from these processes may mobilise at grassroots level, knowing that 'normal' challenges are closed off' (ibid.). As political opportunities evolved on various levels during this period, external political factors, such as economic growth, wider North–South cooperation, EU legislation and the formation of the EPA all influenced the events surrounding these cases.

The apparent disregard by authorities for communities in both jurisdictions in Ireland can be seen as the primary contributing factor (and political opportunity) for environmental movement responses to policy. These various movements shared certain features, and drew from similar pools of internal and external resources, as an anti-multinational and NIMBY frame in the 1970s and 1980s gave way to concerns about the negative impacts of the economic boom in the 'waste management' frame of the 1990s and 2000. GSE's own campaigns were able to use these internal and external resources, as the shifting dynamic of political opportunities evolved, and 'influential allies' were available. The volatile pattern of political opportunities also provided 'unstable alignments' (Tarrow 1994, pp. 85–89) as the state's own legislative framework on waste issues remained concentrated on profitability rather than sustainability, and provided local successes for regional campaigns. As the Irish state's aim of eight regional incinerators has now been reduced to just three (in Dublin, Meath and Cork), local 'success' has been offset by these revised plans for larger incinerators in urban areas.

The political outcomes for individual campaigners have been equally mixed. In 2006, the Green parties in both jurisdictions were unified. In 2007, the Greens had their first representative elected to the Assembly in Belfast. Prominent southern Green party members had come to the fore in local anti-incinerator campaigns. The Greens took the ultimate step towards increasing their access to power after the 2007 general election by entering a coalition government with Fianna Fáil (Leonard 2008a, b). The emergence of the Green party as a coalition partner has had mixed results in relation to incineration, with disagreement about three of the originally proposed eight incinerators continuing. Their combination of framing approaches, combined with the utilisation of EU directives as the basis for legal challenges, represented some success for community groups opposed to the state's waste policy. While plans for incinerators in Dublin (Poolbeg), Meath (with planning permission announced in July 2008) and Cork are still part of state policy, campaigners have continued to oppose their introduction. The Irish Greens have been criticised for their participation in a coalition government which has continued attempts to introduce incineration.

Events in Northern Ireland surrounding the Du Pont campaign seem to vindicate the findings from interviews and focus groups that one of the clearest problems there 'is the democratic deficit in terms of the participation in the construction and implementation of policy' (Fagan *et al.* 2001, p. x). Fagan *et al.* advocate genuine forms of consultation and participation at all stages of the waste management process that might reduce particularistic local reactions against state policies and practices. This was, in part, a response to views such as those found among local authorities that tended to see the issue of waste management in terms of how to avert local opposition to incineration through financial inducements and by *bypassing* local councillors. This may be a self-defeating approach given that research has shown that the type of 'environmental empowerment' necessary for good waste management practice occurs when community control over environmental events exists. Northern Ireland's government has also continued to develop plans for an incinerator in the Belfast region, with environmental groups remaining vigilant.

These case studies demonstrate the complexities surrounding the issue of state or corporate plans for dealing with an ongoing waste crisis in the face of local community concerns about the environment. We have demonstrated that from a governance perspective, the dichotomies between centralised forms of power and community-based movements derived from local political alliances indicate the need for local consultation at the planning and implementation level. While this may delay the introduction of critical infrastructure required by existing levels of consumption, the benefits for the wider political process would include an increased sense of confidence in the processes of development in regions that, for whatever reason, may feel marginalised from the central planning process. The responses of community groups in both jurisdictions can also be seen to have impacted on local politics, and provided an outlet for concerns about wider issues such as democratic deficit. The prevalence of community responses indicates the extent to which degrees of populist local governmentality are manifested in Ireland's body politic in both the North and the Republic. Ultimately, this embedded form of community politics creates an initial but significant layer of civil society which provides an outlet for local political concerns about corporate or state activity, empowers local actors, instigates local political careers and augments local flows of knowledge and networking on issues such as health, politics and regional planning, thereby enhancing the political opportunity framework in a manner that goes beyond NIMBY concerns with the development of regional repertories of power.

While such power is by its very nature transient and fleeting, the temporary leverage it provides creates moments of access for community activists in a manner which is rarely replicated. As such, the significance of local campaigns such as those described here becomes apparent, as local movements can be seen to make an important contribution to the development of community values, meanings and identities.

Notes

1. Manlio Cinalli (2002) has drawn our attention to some interesting specifics of environmental politics in divided societies.
2. The latter stood at 19% in early 2009, and the former at 34%.

References

Allen, R., 1992. *Waste Not Want Not*. London: Earthscan.

Bache, I. and Flinders, M., 2005. *Multi-level governance*. Oxford University Press.

Bevir, M., 2008. The Westminster model, governance and judicial reform. *Parliamentary Affairs*, 61 (4), 559–577.

Cinalli, M., 2002. Environmental campaigns and socio-political cleavages in divided societies. *Environmental Politics*, 11 (1), 163–171.

Davies, A., 2007. A wasted opportunity: civil society and waste management in Ireland. *Environmental Politics*, 16 (1), 52–72.

European Commision, 1989. *A community strategy for waste management communication from the Commission to the Council and to Parliament*. Brussels: Commission of the European Communities.

Fagan, G.H., 2003. Sociological reflections on governing waste. *Irish Journal of Sociology*, 12 (1), 67–85.

Fagan, G.H., et al., 2001. *2001 Waste management strategy: border perspective*. Maynooth: National Institute for Regional and Spatial Analysis.

Forfás, 2001. *Forfás: Ireland National Policy Advisory Body Annual Report 2001*. Dublin: Forfás.

Garavan, M., 2007. Resisting the costs of 'development': local environmental activism in Ireland. *Environmental Politics*, 16 (5), 844–863.

Gaynor, N., 2008. Transforming participation? Unpublished thesis (PhD). Maynooth: NUI Maynooth Library.

Grove-White, R., 1993. Environmentalism: A New Moral Discourse for Technological Society. *In*: Kay Milton, ed., *1993 Environmentalism: The View from Anthropology*. London: Routledge.

Jordan, J. and Gilbert, N., 1999. Think local – act global: Discourses of environment and local protest. *In*: S. Fairweather, ed., *Environmental futures*. Basingstoke: Macmillan, 39–53.

Leonard, L., 2005. *Politics inflamed: GSE and the Campaign Against Incineration in Ireland*. Ecopolitics Series Volume One. Galway: Greenhouse Press with Choice Publishing.

Leonard, L., 2008a. The Irish Greens in the 2007 general election: dealing with the devil or playing for power. *Environmental Politics*, 17 (1), 126–130.

Leonard, L., 2008b. *The Environmental Movement in Ireland*. Dordrecht: Springer.

Murray, M., 2006. Multi-level governance and waste management: the politics of municipal incineration'. *Economic and Social Review*, 37 (3), 447–465.

Offe, C., 1987. Changing boundaries of institutional social movements since the 1960's. *In*: C.S. Maier, ed. *Changing boundaries of the political*. Cambridge UP.

Peace, A., 1997. *A time of reckoning: the politics of discourse in rural Ireland.* London: Routledge.

Rootes, C., 2007. Acting locally: the character, contexts and significance of local environmental movements. *Environmental Politics*, 16 (5), 722–741.

Scott, A., 1990. *Ideology and the new social movements.* London: Unwin Hyman.

Stoker, G., 1998. Governance as Theory: Five Propositions. *International Social Science Journal*, 50 (1), 17–28.

Szasz, A., 1994. *Eco-populism: toxic waste and the movement for environmental justice.* London: UCL Press.

Tarrow, S., 1994. *Power in movement: social movements, collective action and politics.* New York: Cambridge University Press.

Tarrow, S., 1998. *Power in movement: social movements and contentious politics.* 2nd ed. Cambridge University Press.

Wasting energy? Campaigns against waste-to-energy sites in France

Darren McCauley

School of Politics, International Studies and Philosophy, Queens University Belfast

The proliferation of waste incinerators in France has stimulated widespread policy conflict and social mobilisation. French government has struggled to follow a coherent waste management strategy. Its initial preference for landfills as a waste disposal solution was replaced by waste incineration. This sparked a nationwide programme of construction in both urban and rural areas. Recent technological advances in waste-to-energy have further cemented incineration as a major policy solution in its management mix. Local anti-incinerator campaigns have emerged throughout France in response. A resource-opportunity analysis is employed to explain why some campaigns succeeded when others did not.

The development of economically, environmentally and socially effective waste disposal strategies represents a major challenge for society. Each country has built up a particular approach to solving the problem of increasing levels in household, municipal, industrial and toxic waste. Various forms of infrastructure have emerged as a direct response to 'sustainable' waste management. I focus here on household and municipal waste incineration as a prevalent form of waste disposal in France. New technological advances in this area have further promoted incineration through 'energy capture' – the now often termed *waste-to-energy* sites. However, the sustainability of this 'new' waste infrastructure is currently threatened by its inability to successfully allay social anxiety over health, siting and broader justice concerns. I explore recent social unrest in the form of anti-incineration campaigns in France. In particular, I seek to understand why some campaigns appear to succeed in their objectives, while others stagnate.

The incineration issue has provoked much interest from policy-makers in France. Waste management policy has followed a succession of national and

European legislation. A notable characteristic in its development is the early (in comparison to the UK and Ireland) rejection of landfills as the major policy solution. The decentralisation of waste management in 1992 resulted in the proliferation of small-scale incinerators throughout France (reaching 300 sites in 1997). Governmental promotion of recycling in 1998 has led to a largely unique waste management mix shared equally between landfill, recycling and the 'new' waste-to-energy incinerators. The national picture now shows a messy co-existence of new (or renovated) larger urban incinerators and existing small-scale sites in rural/semi-rural France.

Local campaigns have emerged as a direct result of sub-national authorities' pursuing waste-to-energy as a solution to waste disposal problems. The progressive decentralisation of waste management has, therefore, provoked instances of social unrest throughout France. I concentrate on two contrasting examples: one of a successful campaign (in Coulon North-West France); the other of a stagnated campaign (in Clermont-Ferrand, Central France) against proposed waste-to-energy sites. Similar research on social movements in France has underlined the changing nature of environmental activism in terms of action repertoires (Hayes 2006), networking (Guigni 2001, Duriez 2004), ideology (Beroud *et al.* 1998) and influence (Bechmann 2002). Moreover, it has revealed a mixed picture of an overbearing state apparatus (van der Heijden 1997, Dryzek *et al.* 2003) under pressure from decentralisation and Europeanisation (Hayes 2002, Waters 2003).

Scholars in social movement research have equally focused attention on the particular characteristics of anti-incinerator campaigns. There have been several works dedicated to the examination of ecological discourse explanations in research conducted on anti-incineration movements (Walsh 1988, Kubal 1998, Saarikoski 2006). European literature in this area has especially focused on discourse/framing analysis (in addition to other conceptual tools) in anti-incinerator campaigns (Hunter and Leyden 1995, Davies 2005, 2007, 2008, Leonard 2006, Rootes 2006). I contribute to this literature through an examination of resources as a potential explanatory variable in campaign outcomes. As do Leonard (2006) and Rootes (2006), I undertake an alternative analysis of political opportunity structures as a competing set of explanations.

Waste management in France: from landfills to waste-to-energy

Incineration emerged in France as a major policy option for waste reduction, and in particular, as the major alternative to landfill. The early (1970–1990) development of waste management in France is marked by the official rejection of landfills as the main policy solution. Following trends throughout Europe, waste management policy emerged in the 1970s with the 1975 Waste Act and the establishment of the National Waste Recovery and Disposal Agency (Agence national pour la récupération et l'élimination des déchets). Three forms of landfill sites were prevalent throughout this period: class 1 (hazardous waste), class 2 (industrial or household waste) and class 3 (inert mining or building waste). As the cheapest form of waste disposal, landfill sites became overused and ultimately contaminated water tables (Mathieu 1992).

Two scandals provoked public outcry against the use of landfills. Barrels of dioxin were, firstly, imported into France and illegally dumped in Roumanzières in 1983 after an explosion at a factory in Milan. Secondly, inadequate safeguards were put in place over the disposal of hazardous wastes in Montchanin in 1985 (Mathieu 1992). The 1990 Lalonde 'Green Plan' established a proximity principle whereby waste must be treated as near to its source as possible. It stipulated that each local authority (département) could only have one class 2 (industrial or household waste) landfill site. As a result of public scandals and legislation, 10,000 landfill sites were closed between 1978 and 1990 (Bertolini 1998).[1]

The 1992 Waste Act (transposing EC directives 89/396 and 91/156) decentralised waste management through a clear delineation of powers where local (département) and regional authorities managed household waste and industrial waste respectively. These new powers were accompanied by significant fines for improper disposal. Post-1992, the préfet or the conseil général are obligated to draw up waste strategies at local level (département). They establish recycling objectives, plans for the modernisation of waste treatment and corresponding investment. Any change to local waste management must adhere to the overall strategy developed by the préfet or the conseil général. The first local management strategies placed too much emphasis on replacing landfill with outdated forms of waste incineration (Bertolini 1998, Szarka 2002).

The drive for greater efficiency, up-to-date technological processes and the reduction of emissions has dominated the development of waste incineration policy. Three separate phases are apparent in the development of French policy on waste incineration: large-scale investment in incineration (1992–1998); scaling back of the incineration programme (1998–2003); the application of new technology to reduce harmful emissions (2003+). The 1992 Waste Act set in motion a policy aimed increasing the use of recovery, recycling and energy conversion. Local authorities throughout France constructed new energy recovery incinerators that produced electricity and heating. Due to the large investment required by authorities, management strategies prioritised an increase in tax revenue alongside large-scale public–private partnerships. Budget requirements dictated the construction of notably small-scale energy recovery incinerators.

In 1998, the Environment Ministry circulated advice to local authorities emphasising the need to promote dialogue and recycling. The government was concerned by the de facto privatisation of the incineration programme due to a significant shortfall in public investment. Public perception was influenced when three incinerators near Lille were closed down in 1998 after dioxin was found in the milk of nearby cows (Szarka 2002, p. 181). EU waste legislation recently raised cautious opposition to incineration as potentially harmful to the environment in the Waste Incineration (WI) directive 2000/76/EC and the

Waste Framework (WF) directive 2006/12/EC. In France, a 2003 law transposing the WI directive concentrated on the promotion of new technological processes (especially waste-to-energy) alongside further reductions in emissions. It equally marked the official closure of out-of-date waste incinerators not conforming to EU specifications.

Key figures

I provide below the key statistics on waste incinerators in terms of their number, age, capacity and position in the overall waste management mix. Before delving into more comprehensive data, Figure 1 shows that there are more waste incineration plants in France than any other Western country (where data is available) (140 in 2001 and 129 in 2007) even though these figures refer to the period after a massive reduction in out-of-date incinerators from 300 to 140 between 1998 and 2001 (explored below). France's closest rival, Germany, has just over half the number of incinerator plants (66 in 2007). The combined total of incinerators in France and Germany is larger than that of the remaining 13 countries in focus. Since 2001, there has been relatively little change in the number of incineration plants throughout Europe. There have been modest reductions in Belgium, Denmark and France, while incineration plants have maintained or increased their presence in the remaining countries.

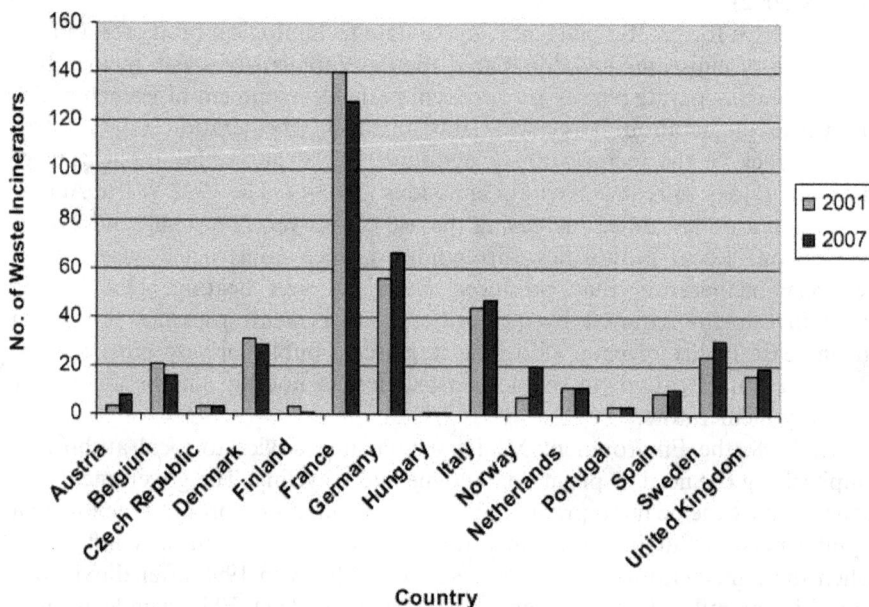

Figure 1.   Number of waste incinerator plants in Europe. *Source*: European Waste to Energy Plants, (ISWA) Reports 2001 and 2007.

However, the International Solid Waste Association (ISWA) and Eurostat reports also highlight that France treated significantly less waste in 2007 by incineration than Germany – 12.3 million tonnes in comparison to 17.4 million tonnes – in spite of its superior count in incineration plants (ISWA 2001, 2007, Kloek and Jordan 2005). Brousse (2005) explains that France has significantly *reduced* its treatment of waste via incineration, and resulting dioxin emission levels, since 2000. These statistics reinforce the local management structure of waste incineration in France post-1992. The prioritisation of waste incineration by local authorities resulted in the construction of many small-scale incinerators. Moreover, the 1998 review and the 2003 transposition of directive 2000/76 clearly provoked a reduction in the number of incinerators. A closer inspection at the French case in European perspective is needed to explore figures on the overall waste management mix.

The waste management mix in France (see Figure 2) reveals an unusually close share between landfill (35%), waste-to-energy (33%) and recycling (32%). This mix contrasts sharply with Ireland, Spain and the UK where there is a high dependency on landfill. The shifting prioritisation from landfill to waste-to-energy with increasing emphasis on recycling (post-1998) explains this. France has placed itself as a European leader (alongside Denmark, Sweden, Belgium and Luxembourg) in waste-to-energy technology. The high level of waste treatment by waste-to-energy technology in Germany (above) is dwarfed by recycling as its main policy solution (24% vs. 59%). Moreover, the output from waste-to-energy sites in France provides the second most important renewable energy source for electricity (after water power) and heating (after wood) at a total annual rate of 12,000 GWh (Gigawatt hours). Out of 13 million tons (85% of which comes from households), over 95% of waste incinerated uses state-of-the-art (conforming to EC law) energy recovery technology (DGEMP 2005).

Taking a closer look at France, the promotion of new waste-to-energy technology in the 1998 review led to the closure of over 160 out-of-date incinerators (from a total of 300 in 1998). The small-scale incinerators developed by local authorities were replaced by fewer larger sites (85% in partnership with the water and waste company, Suez Environnement) that conformed to national and European specifications. Indeed, Brousse (2005) argues that fear of waste incineration plants in France is misplaced as emission levels are now strictly regulated under EC legislation. There were, in May 2008, 129 active waste incineration plants. Their location is notably concentrated in urban centres, especially in and around Paris, Bordeaux, Marseille, Lyon and Lille (accounting for 56 in total). The three most active localities (measured by emission levels) in France mirror the urbanisation pattern: Paris, Seine Saint Denis and Rhône.

Table 1 reveals the most active waste-to-energy sites in urban areas in France (as measured by the average annual waste incinerated). Ivry, placed centrally in Paris, is the most active site in France, with a capacity of 100 tons per hour. However, the combined annual output of the five sites only represents a tenth of the national total. The majority of active incineration

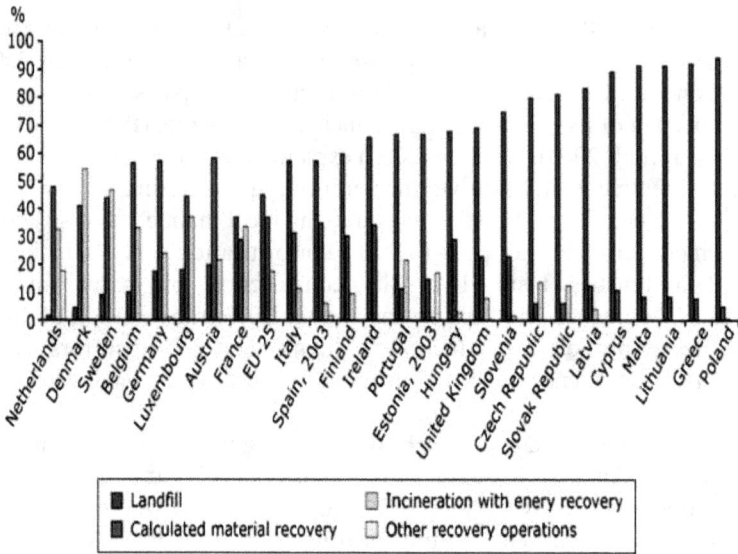

Figure 2.   Data on the waste management mix in Europe 2006. *Source*: Eurostat (2006) Municipal waste by type of treatment.

Table 1.   The most active incinerators in urban areas in France.

Location (in vicinity of)	Main incinerator	Established (renovated)	Average annual waste incinerated (tons)
Paris	Ivry	1969 (1997)	699,250
Lille	Haulin	1998	350,000
Bordeaux	Bègles	1998	255,000
Lyon	Lyon Sud	1989 (2002)	235,000
Marseille	Lunel-Viel	1999	120,000
Total average annual waste incinerated in France			13,882,120

Source: MEDD (2008a).

plants (73 in total) are still located outside these urban areas. Additionally, 68 of the 81 incinerators decommissioned since 2000 were situated outside the urban centres of Paris, Bordeaux, Marseille, Lyon and Lille (MEDD (Ministere de l'Environnement et de Developpement Durable) 2008a, pp. 4–9). All 12 incinerators currently under proposal are also found in small cities, towns and rural areas (MEDD 2008b, p. 3). For that reason, I concentrate here on proposed waste incineration plants in rural and semi-rural/urban France.

Exploring the outcomes of anti-incinerator campaigns

There is insufficient data to map all the anti-incinerator campaigns throughout France. My focus here is, therefore, on the most high-profile campaigns (in terms of media coverage). Since 2001, there have been seven high-profile campaigns that have resulted in the abandonment of proposed incinerators. In the South of France protesters managed to prevent the establishment of an incinerator in the small Mediterranean town of Vias near Montpellier in 2005 (Martin and Olivier 2005). In the north-east, a local association, 'Vigilance Projet Incinérateur Gueugnon', energised by its charismatic leader, Alain Rault, was at the heart of local efforts to reject an incinerator at Gueugnon in 2003 (Jacques 2003). In the west, four proposed incineration plants (St. Capraise, Izon, Grosbreuil and Angers) were officially abandoned between February 2002 and December 2006 as a result of local campaigns from a range of both institutional and non-institutional actors (MEDD 2008a, p. 10).

I concentrate on one ultimately successful campaign at Coulon in Poitou-Charentes (Western France) and one protracted campaign at Clermont-Ferrand in Auvergne (Central France). I based case study selection on three criteria. Both cases are, firstly, located outside the main urban basins of Paris, Bordeaux, Marseille, Lyon and Lille. Secondly, Coulon, a small town surrounded by a major national park (le Parc du Marais Poitevin), is an appropriate representation of rural France. Clermont-Ferrand is a major regional city in relative isolation from other major urban centres. Lastly, the success experienced by the anti-incinerator movement in Coulon took place throughout a short duration of only three years (2003–2006). At Clermont-Ferrand, campaign activity against a proposed incinerator has continued since 2001. I conducted documentary research alongside semi-structured interviews in both areas.

How do we explain both the success of the Coulon anti-incinerator campaign and the apparent stagnation experienced at Clermont-Ferrand? Social movement theory (broadly defined) provides a framework for studying group behaviour while allowing us to generate questions about how and why social mobilisation takes place. It is not reasonable, or necessary, to discuss the complexities of various social movement theories in depth as there is a plethora of such works (Bell 2001, Kriesi 2004). However, it is useful to identify such debates in research into mobilisation against waste infrastructure that might assist in the development of a conceptual framework for understanding campaign outcomes at Coulon and Clermont-Ferrand.

The first approach identified within literature on anti-incinerator campaigns is ecological discourse analysis. As the most prevalent theoretical instrument in this area, ecological discourse analysis is used to emphasise the role of identity and ideology in anti-incinerator campaigns. It is closely related to the examination of collective framing processes. Framing analysis has been employed as a means to dismiss NIMBYist stereotypes of anti-incinerator campaigns (Walsh 1988, Hunter and Leyden 1995, Kubal 1998). Gerrard (1996) and Visiglio and Whitelaw (2003) reveal the close link between ideas of environmental justice and controversy over the siting of waste incineration facilities. Research conducted in Finland also suggested that 'framing conflicts' resulted in an initially ineffective anti-incinerator movement (Saarikoski 2006).

Building upon framing processes, Davies (2005, 2007, 2008) introduced the notion of 'cultures of action' as a means to exploring the interconnected explanatory power of culture (feelings and practices).

I concentrate here on a qualitative approach to 'resource mobilisation theory' and 'political opportunity structures' as a theoretical approach that, combined, has been under-employed in European literature on anti-incinerator campaigns. Research conducted in the US explored a resource mobilisation explanation to the outcomes of anti-incinerator campaigns based upon the role of expertise, public support and financial backing within urban and sub-urban areas (Walsh *et al.* 1997). Leonard (2006) integrates resource mobilisation analysis into his research on the Irish environmental movement, his study of anti-incinerator campaigns in Galway, Meath and Cork concentrating rather on discourse/framing and examination of political opportunities. Adopting a similar theoretical approach, Rootes (2006) concludes from his research into several anti-incinerator campaigns in England that political opportunity structures proved more decisive than ecological discourse in explaining outcomes. I follow an alternative comprehensive analysis of mobilisation outcomes through a combined resource-opportunity examination.

The resource argument

A resource analysis emphasises the key role of 'rational incentives for collective action' (Appleton 2000, p. 39). Firstly, the size difference between Coulon and Clermont-Ferrand is a potential explanatory variable for campaign outcome. Coulon is a small town with only 2174 inhabitants in the north-west region of Poitou-Charentes. The local economy is heavily based on non-intensive agriculture and tourism. Known as the capital of 'Green Venice', it is an established venue attracting tourists from throughout France. Any potential population growth is restricted by its location deep within the Marais Poitevin, a 100,000 hectare protected area that stretches across three local administrative areas (départements) – Charente-Maritime, Deux-Sèvres and the Vendée. Waste disposal is managed by the larger (58,066 inhabitants) regional city of Niort (INSEE 2006).

In stark contrast, Clermont-Ferrand, with a population of 140,000 inhabitants, is over 70 times larger than Coulon. It is the capital of the Auvergne region in southern France overlooking the plain of Limagne in the Massif Central. The local economy has historically depended upon the manufacturing industry and in particular the tyre manufacturer, Michelin (since 1910). It also hosts the regional headquarters for finance and banking, administration and transport (INSEE 2006). Consequently, regional waste management strategies are developed in Clermont-Ferrand. From a resource perspective, a campaign in Clermont-Ferrand should benefit from the increased access to experts, financial assistance, transport infrastructure and coalition partners in a large city (Walsh *et al.* 1997, Tarrow 1998). Yet the anti-incinerator campaign at the small town of Coulon was more successful over a shorter time (2003–2006). In order to provide a more comprehensive analysis, it is necessary at this point to briefly examine differences between the resource bases of anti-incinerator campaigners and proponents at Coulon and Clermont-Ferrand.

From a campaigner's perspective, resource mobilisation theory posits simply that adequate levels of resources are needed for a successful mobilisation on a given issue. In other words, a movement actor 'can do no more than its resources … permit' (Freeman 1979, p. 167). A campaign to avoid the construction of an incinerator at Coulon emerged in early 2005. Its participants included Les Verts (Green Party) and local environmental association DSNE – 'Deux-Sèvres Nature Environnement' – a constituent member of the national federation France Nature Environnement. Most notably, the leader of the campaign (Patrick Morin) described himself as an 'unaffiliated ordinary citizen' (Martin 2005a). In March 2005, a newly formed anti-incinerator committee of *Coulon* organised a protest with demonstrators draped in orange[2] (Baroux 2005). The inclusion of a new participant, 'Citoyens pour l'Information dans le Mellois sur l'Environnement et la Santé' (CIMES), in the committee consolidated the role of Les Verts, DSNE and Patrick Morin as the principal campaign actors.

The campaign at Clermont-Ferrand involved significantly more civil society actors. In 2003, an 800-strong demonstration at the city hall marked the founding of the main protest confederation, le Collectif Contre l'Implantation à Proximité de l'Agglomération Clermontoise (CCIPAC).[3] In addition to combining their own financial clout, the 12 regional associations brought together in CCIPAC were able to exploit the resource base of their smaller subsidiary groups throughout Auvergne. Moreover, this highly organised confederation was supported by notable political backing. In 2004, the anti-incinerator campaign was reformed under a protest confederation of public figures (mayors, civil servants and scientific experts) and the associations of CCIPAC known as PROPRE.[4] Deeper analysis of individual campaign resources is not warranted as the better equipped Clermont-Ferrand anti-incinerator movement enjoyed *less* success than its Coulon counterpart.

It is difficult to measure differences in public support as a resource for the campaigns at Coulon and Clermont-Ferrand (no comparable polling data are available). Mobilisation activities were logically on a smaller scale in the former case. In 2005, a petition against the proposed incinerator was lodged with initially 400 signatures (rising to 800 by March). The most significant demonstration gathered together 1500 activists in Coulon with placards proclaiming 'Dioxin City' (Biard 2005a). With regard to Clermont-Ferrand, a public consultation took place in 2003 in the suburban town of Lempdes. In a referendum on the incinerator at Clermont-Ferrand, turnout was 41.2%, with 96% voting against (CCIIPAC 2003). In terms of mobilisation, between 2500 and 3500 demonstrators filled the streets of Clermont-Ferrand in March and November 2006 while a further 30 campaigners emptied their bins in front of city hall in August (Robert 2006).

The resource base of the proponents of incineration can also be a decisive explanatory variable (Leonard 2006). The main institutional actor involved in promoting the idea of an incinerator at Coulon was the Communauté d'Agglomération de Niort (CAN). Representing 100,000 inhabitants in 29 local authorities (communes) treating 47,000 tons of waste per annum, it was first established (by the regional authority) in 1999 to represent the locality surrounding the main city of Niort. This publicly elected body is chiefly responsible for land and services (including waste) management in the localities it represents (Moinet 2007, pp. 3–5). In September 2003, it commissioned a report on the future of waste management in Niort and surrounding areas.

Accepting its proposals, the CAN announced in December 2004 its preference for incineration instead of landfills, with the precise location undecided at this point (Lyvinec 2004). The main actor involved in the pro-motion of an incinerator was a publicly elected body, VALTOM (Valorisation et de Traitement des Ordures Ménagères) which represents 654,000 inhabitants in over 544 local administrative units (communes), almost seven times more than CAN, and has an operational turnover of 362,900 tons of municipal waste. Its membership includes trade unions, business and delegates as political representatives (local mayors and civil servants). In January 2003, VALTOM officially announced that an incinerator would be completed on the Clermont-Ferrand site within three years.

There are three important conclusions in understanding the differing resources of the two proponents of incineration and explaining the outcomes of the contention. Firstly, VALTOM has, unsurprisingly, competence over a more substantial waste management strategy with higher waste tonnages to process. Secondly, CAN manages waste disposal for another 28 localities – communes – in addition to Coulon whereas VALTOM is more specifically concerned with Clermont-Ferrand and its immediate surroundings. CAN is a regional authority that, thirdly, develops policy on a number of areas (economic development, public transport, education). VALTOM is a public body that can dedicate all its resources to waste management in the Clermont-Ferrand area. These differences underline that VALTOM was better resourced to counter a campaign against the proposed incinerator than was CAN.

Political opportunity structures: from waste to biodiversity structures

Political opportunity structures provide a framework to examine the macro-level of resource mobilisation arguments. Collective action involves rational actors that attempt to realise certain objectives within an ever-changing larger political environment (Clark 2002, p. 414). Before 1992, central state agencies developed waste management strategies in accordance with a national plan of action. Opportunity for civil society involvement was extremely limited. The decentralisation of waste policy in the 1990 Lalonde Green Plan offered non-governmental actors a new venue for action in implementation and decision-making processes. However, the period 1992–1998 revealed that only business enjoyed any increased formal access to policy-making as local authorities focused on public–private partnerships (Szarka 2002).

The waste management companies, Société Industrielle des Transports Automobiles (SITA) and Lyonnaise des Eaux (now owned by Novergie) developed a programme for the construction of small-scale incinerators with local authorities (as well as regional authorities in respect of industrial waste) (Bertolini 1998). The review conducted in 1998 by the Environment Ministry underlined the need for more dialogue with societal actors in local waste management strategies. As a direct result, 'Local Waste Contracts' (Contrats Territoriaux Déchets) emerged as a formal mechanism for including civil society organisations in policy decisions. The environmental agency ADEME (l'Agence de l'environnement et de la maîtrise de l'énergie) manages the scheme nationally, closely coordinating with local authorities (département). These contracts are signed by local government, business and civil society actors.

Their objectives are established in accordance with the préfet in order to achieve a more coherent and publicly supported waste management strategy, with varying levels of funding attached. There are currently 112 contracts either in process or fully completed throughout France (Lepellier 2008). In the case of Coulon, a contractual agreement was only reached after the proposed incinerator failed in 2006 (ADEME 2008). The funding attached to the contract may have been an important incentive for reaching an agreement on the incineration issue, but any associated opportunity structures could not have influenced the campaign's outcome as they were established after its completion. Clermont-Ferrand has failed to agree upon any such contractual arrangement. The unresolved incineration issue has hampered the development of any consensual waste management strategy.

Biodiversity issues in waste incineration

I argue below that opportunity structures developed in biodiversity policy are, in contrast to waste management structures, influential in explaining the different outcomes of the Coulon and Clermont-Ferrand anti-incinerator campaigns. As demonstrated below, both proposed incineration plants were located near zones for special protection (ZSPs) under the EU N2000 programme. Of course, these areas of natural beauty existed before the establishment of any European programme. The unique aspect contributed by the EU N2000 programme was the necessity to introduce new decision-making structures and processes at each ZSP. Article 6 (1) of the Habitats Directive 92/43/EEC states that '[f]or special areas of conservation, Member States shall establish appropriate management plans specifically designed for the sites and appropriate statutory, administrative or contractual measures'. In France, local contracts between stakeholders emerged alongside the establishment of steering committees as a significant opportunity structure for civil society actors.

Figure 3 shows the scale of the park covering 68,023 hectares (ha) to the west of Niort. It is the biggest site in Western France, and the small town

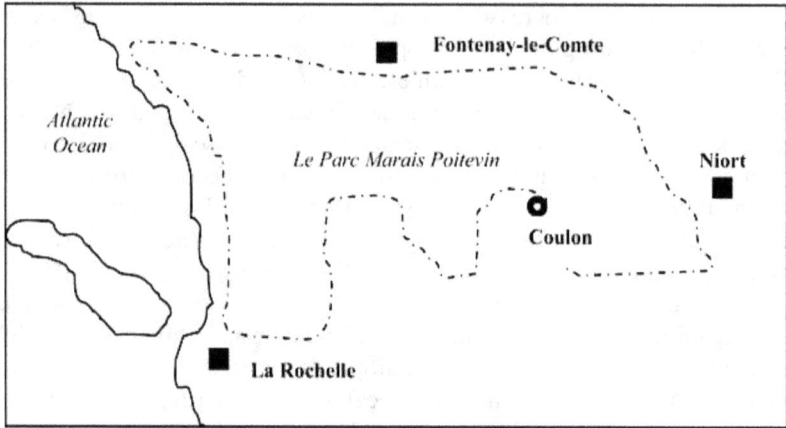

Figure 3.   Zones for special protection at Coulon: Le Parc Marais Poitevin.

of Coulon (black circle on Figure 3) is effectively surrounded by parkland. The anti-incinerator campaign focused therefore on the potential threat of dangerous emissions to wildlife and disruption of local habitats found in the park (CIMES 2005). Although France communicated to the European Commission its willingness to include the Marais Poitevin as a ZSP in 1996, it was only in 2001 that a steering committee was eventually formed in order to prepare for its status as a ZSP. Its official establishment coincided with the first CAN report into the possibility of an incinerator at Coulon. In 2003, the committee became fully operational with agreement upon its structure and primary objectives over the next six years (under the so-called 'documents d'objectifs' [Docob]) (DIREN 2003).

In contrast to the Coulon anti-incinerator campaign, the three recently established ZSP steering committees at Clermont-Ferrand exerted compara-tively *less* influence on the development of its anti-incinerator campaign. The proposed site is located deep within a valley, a designated ZSP itself (see Figure 4). The valley protects a wide range of rare species and habitats over a relatively small area (231 ha) (European Commission 2006b). Immediately south of Clermont-Ferrand, the second ZSP is a much larger area (51,853 ha) with protected riverland, forest and associated habitats (European Commission 2006a). Both achieved full status under the European N2000 programme in early 2006. The third ZSP (la Chaine des Puys) represents the most famous (and controversial in terms of anti-incinerator demonstrations) site. Directly west of the city and the proposed incinerator lies the most expansive volcanic range in Europe, home to the Puy de Dôme volcano. It achieved full recognition under N2000 in 2005 (European Commission 2005a). However, the steering committees in all three cases are yet to become fully operational.

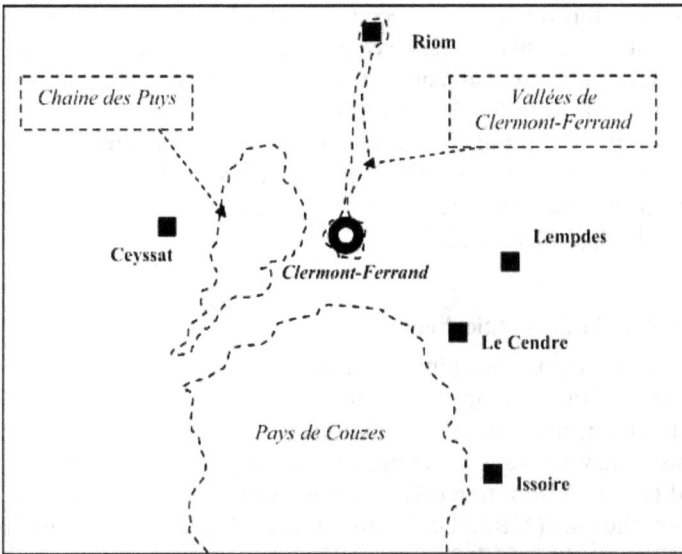

Figure 4.   Three zones for special protection at Clermont-Ferrand.

Biodiversity steering committees and anti-incinerator actors

N2000 resulted in the establishment of regional-level contracts alongside a powerful financial instrument called LIFE-Nature (McCauley 2008). New forms of local-level institutionalised interaction between stakeholders (so called 'comités de pilotages' – steering committees) are integral components in a contractually agreed set of objectives ('document d'objectifs' or Docob) (Le Grand 2004). These steering committees became involved in disputes over planned waste incinerators near the relevant ZSP. Indeed, the Habitats Directive 92/43/EEC in Article 6 (3) states that 'any plan or project not directly connected with the management of the site but likely to have a significant effect thereon shall be subject to appropriate assessment of its implications for the site'. I explore below the different ways in which steering committees at ZSPs in close proximity to proposed incineration plants can influence opportunities for campaigns.

In the same year (2004) that the CAN report on an incinerator for Coulon was published, the Parc Marais Poitevin was already officially recognised as a ZSP under European law (European Commission 2005b). A fully operational steering committee (since 2002) included the regional authority of Poitou-Charentes, regional council representatives, environmental, forestry, agriculture and water ministerial agencies. In terms of non-state actors, there were representatives from environmental, hunting and agricultural associations as well as the scientific community. Crucially, a main objective in the agreed 'Docob' was to 'evaluate the state of habitat conservation and identify human activities likely to help as well as hinder its longevity' (DIREN 2003, p. 3)[5]. The committee used this objective to become involved in disputes over a proposed incinerator in its vicinity. It is revealed below that this steering committee played a significant role in shifting opportunities for the anti-incinerator campaign. In contrast to Coulon, it is argued that the steering committees in all three ZSPs in the Clermont-Ferrand area were comparatively less involved in the anti-incinerator campaign. All three committees have yet to agree upon a set of objectives (and therefore are not yet operational) as the sites were only recently established (2005 and 2006).[6]

Splits and allies in the political elite

The eventual divergence of opinions expressed by political actors was integral to the demise of the Coulon incinerator project. The steering committee was itself influential in promoting splits among the political elite. Represented on the Marais Poitevin steering committee, the mayor of Bênet (Daniel David) responded to the decision to locate an incinerator in Coulon as a 'choice that is contrary to the NATURA 2000 project and the preservation of the Parc Marais Poitevin'[7] (Martin 2005b). His office underlines that its participation in the steering committee was influential in his hostile stance against the proposed incinerator (Interview 13 February 2006). In a high-profile dispute, Daniel David publicly criticised the mayor of Aiffres (Alain Mathieu), who supported the incinerator at Coulon – 'Alain Mathieu is contributing to the disfiguration of the Marais Poitevin'[8] (Riand 2005). As both mayors represented the same party (Socialist Party), it is David's involvement in the steering committee that appears to have been influential in his opposition to the incinerator.

Another split in the Socialist Party appeared when the newly created presidency of the Marais Poitevin park was awarded in 2004 to Ségolène Royal, the future Socialist candidate for the Presidency of the French Republic. In a largely ceremonial position, she presided over the annual meeting of steering committee members. Refusing to criticise Alain Mathieu as the primary supporter of the incinerator,[9] she declared that 'the incinerator plant was outside her area of competence' (Morel 2005, p. 6). On 1 February 2005, Royal met with the steering committee for the first time (Interview 30 August 2006b). Later that month, she reversed her initial approach, stating that 'I am surprised that such equipment will be installed in the parkland ... it is difficult to ask farmers to adhere to the Natura 2000 framework and then see an incinerator come along'[10] (Girard 2005). Alain Mathieu attempted, without success, to play down the growing splits within the Party, claiming 'this is not about opposition between personalities!' (Lefevre 2005, p. 5). However, the former president of the 'conseil général de la Vendée', Philippe de Villiers, blasted Royal for bringing 'a Parisian view of the Marais', claiming it was outside her competence (Boyer 2006).

There were also splits among the political elite in the Clermont-Ferrand case. Indeed, the founders of the protest confederation PROPRE in 2004 comprised the Mayors of Aulnat (Didier Laville), Cournon (Bertrand

Pasciuoto) and Lempdes (Jean-Pierre Georget) (Sauges 2006). The distinction lies in the contrasting roles played by the ZSP steering committee in each case. There was no evidence for committee influence in Clermont-Ferrand (Interview 13 April 2006).

In addition to gaining the support of political party members (such as Daniel David and Ségolène Royal), the Marais Poitevin steering committee provided more generally an opportunity for anti-incinerator campaigners to discuss their views with state agencies and local authority representatives from Coulon and Niort. In particular, the environmental association Deux-Sèvres Nature Environnement was an integral player in both the campaign and steering committee. The committee convened four times every year in order to set and examine agreed objectives. A representative from Deux-Sèvres Nature Environnement underlined that a lobby group against the proposed incinerator presented its opposition to the préfet in January 2005 (Interview 30 August 2006a). Crucially, the head of the steering committee was a representative of the préfet, Jean-Jacques Brot. The lobby included representatives from environmental, forestry, agriculture and water agencies, landowners, associations and independent scientists (from le Conseil de l'Ordre des Médecins).

In practice, the steering committee became monopolised by the incinerator issue throughout 2005 (convening formally for an unusual six times). A group of 12 scientists involved in the steering committee sent a letter of protest to Brot as head of the committee as well as CAN on 27 January 2005 (CPAI 2005).[11] Deux-Sèvres Nature Environnement withdrew its involvement in the steering committee in protest against the proposed incinerator between May 2005 and January 2006. The environmental association claimed to have informally met with committee members (including the office of the préfet) during this period (Interview 30 August 2006a). The absence of any steering committee at Clermont-Ferrand deprived campaigners there of any similar opportunity to promote splits and allies in the political system.

Discussion

The proliferation of waste-to-energy sites in France has evoked considerable social unrest. Anti-incinerator campaigns at Coulon and Clermont-Ferrand offered contrasting examples of conflict between societal and policy actors. The examination of resources and opportunity structures in both case studies revealed several variables the consideration of which helps to explain the divergent outcomes.

Resource analysis and anti-incineration campaigns

Four key debates on resources emerged from a qualitative analysis of the two cases: *access, network, public support* and *opposition*. A rural–urban division between Coulon and Clermont-Ferrand offered the potential for varying levels of material, human or network resources. In terms of resources, the geographical location of a campaign is vital for its ultimate success or failure (Franquemagne 2007). This debate concentrated on the notion of differing '*access*' to resources (Edwards and McCarthy 2004). Indeed, the existing literature on campaigns against waste incineration tends to focus on either rural (Davies 2007, 2008) or urban (Walsh *et al.* 1997) areas. The campaign at Clermont-Ferrand had better access to transport, manufacturing and above all waste management infrastructure. As the capital of the Auvergne region, policies on waste were developed locally in Clermont-Ferrand, whereas for Coulon they were decided outside the locality. However, greater access to a more substantial infrastructure in Clermont-Ferrand did not result in a more successful campaign, in terms of either duration or of result.

The second debate examined '*network*'-based resources for campaigners in both cases (Guigni 2001, Duriez 2004). Organisations in domestic networks increase their resource base through a mutual exchange process; groups seek to participate in these networks to allow them to fight local battles drawing upon outside and local resources (Tarrow 1998, pp. 187–188). The campaign at Coulon was based on a small network of societal and political actors. The highly organised Clermont-Ferrand network involved more civil society actors, mayors, civil servants and scientific experts. The noticeably larger network (in number and size) appeared to be less successful than its modest counterpart. As in the 'access' debate, an examination of 'network' resources underlined its ineffectiveness in explaining the campaign outcomes. An alternative closely related explanation might involve an analysis of the role of Patrick Morin as a particularly effective 'leader' of the kind often found in small local networks (Coban 2004).

The existing data on '*public support*' proved relatively inconclusive. There was clear evidence for public support in the form of the petition raised at Coulon, and from the results of the referendum concerning the proposed Clermont-Ferrand incinerator held in the suburban town of Lempdes. The most convincing resource debate that emerged above centred on '*opposition*'. There was a notable distinction between how and where the respective pro-incinerator authorities could commit resources. VALTOM, as the major authority promoting incineration, benefited from its ability to focus all its resources on the waste incineration issue and in Clermont-Ferrand. CAN, by contrast, divided its resources among other issues (economic development, public transport, education and waste treatment) across a wider area (the city of Niort and its surroundings). Moreover, the higher tonnages of waste requiring treatment in the Clermont-Ferrand area offered a powerful 'legitimising' resource for those promoting the establishment of a waste incinerator.

Opportunity structures in France: between decentralisation and Europeanisation

The first critical distinction materialised between the political opportunity structures in waste management and those found in biodiversity policy.

The progressive decentralisation of control over waste policy in France led to the establishment of Local Waste Contracts which, in theory, provide an opportunity for all stakeholders to participate in drawing up local waste management strategies. However, the stalemate experienced on the incinerator issue in both cases prevented these contracts from offering any new opportunities for access to policy-making. In contrast to waste policy, the regime surrounding biodiversity policy provided opportunities that bore more strongly upon the outcomes of the campaigns.

The emergence of an influential steering committee at Coulon, unlike Clermont-Ferrand, proved to be critical. Biodiversity policy in France underwent a process of decentralisation similar to that concerning waste, with European legislation on N2000 exerting significant influence (McCauley 2007). In Coulon, a contractually agreed decision-making process was established in 2002 to oversee the protection the Marais Poitevin parkland. The Marais Poitevin steering committee (with significant funding attached to an agreed set of management objectives) became increasingly active on the incineration issue. The committee encouraged splits between local Socialist politicians (Daniel David, Ségolène Royal and Alain Mathieu) as well as providing a venue for allegiances to form between anti-incinerator societal, scientific and governmental representatives.

Anti-incineration campaigns take place within a French state apparatus that is increasingly characterised by the twin pressures of decentralisation and Europeanisation (Hayes 2002). Regional authorities (in respect of industrial waste) and local authorities (with respect to municipal waste) hold significant power in developing waste management strategies and ensuring implementation of national policy. This trend is not restricted to waste disposal. Increasing powers have also been accorded to local government and non-governmental representations in biodiversity policy. In respect of both waste and biodiversity, new opportunities have emerged for societal actors at a sub-national level. However, both policy domains underline the key role played by European legislation in this decentralised opportunity structure. The 1992 Waste Act accorded greater powers to sub-national authorities (leading to Local Waste Contracts) in response to EC directives 89/396 and 91/156. Similarly, the Habitats Directive 92/43 encouraged the establishment of steering committees and contractual agreements in biodiversity policy.

Waste incineration or, in its new guise, waste-to-energy, remains a divisive issue for both authorities and local stakeholders. The French case has revealed a mixed picture in terms of both policy development and social mobilisation. The former has suffered from a consistent lack of direction from national government. The decision to devolve waste management powers to sub-national authorities resulted in the proliferation of small-scale incinerators in both urban and rural areas; the marked reduction in the numbers of incineration plants from 300 to 140 has not continued post-2002. Recent technological advances in waste-to-energy, in line with European legislation, have encouraged the renovation and establishment of large-scale incinerators in urban locations.

In response, societal actors have organised numerous anti-incinerator campaigns throughout France. In light of seven high-profile media successes, the case of Coulon contrasted with the stagnation experienced at Clermont-Ferrand. Nonetheless, both examples revealed active movements comprising of civil society organisations, politicians, civil servants and scientists. An examination of resources showed how differences prevailed in terms of access, networks, public support and in particular the capacities of the proponents of incineration. In addition, an overall pattern of decentralisation and Europeanisation is observable in terms of political opportunity structures. Surprisingly, the key variable explaining divergent campaign outcomes appears to be the opportunity structures established by the regime governing biodiversity rather that than concerning waste.

Acknowledgements

A previous version of this paper was presented in Mannheim, Germany, and more recently, at Trinity College Dublin. The author thanks the contributors for their very helpful remarks. He also thanks the European Commission and UACES for a generous grant for interview and archival research conducted in France and Belgium.

Notes

1. The early rejection of landfills as the major policy solution to waste disposal negated the impact in France of EC directive 1999/31 on the landfill of waste (unlike the UK and Ireland).
2. The decision to wear orange was inspired by the 'orange revolution' in the Ukraine in December 2004.
3. ACIIPAC (Association Contre l'Implantation de l'Incinérateur à Proximité de l'Agglomération Clermontoise), ADEC (Association de Défense de l'Environnement de Chateldon), ADEL (Association de Défense de l'Environnement de Lempdes maison des associations), ARMURE (Vertaizon), Bien être à Aulnat, Brigades Vertes, Comité d'Aménagement de Clermont-Est, FCPE Lempdes (Fédération des Parents d'Elèves de Lempedes), ARB (Association des Riverains de Beaulieu), Gerzat Environnement, VELOS Vélo Environnement Santé, LAVE (Lempdes Association Vie Environnement).
4. PROPRE is an acronym derived from 'Collectif Pour Repenser l'Organisation et le Plan de Recyclage et d'Elimination des déchets'. An interview (30 August 2006b) concluded that business interests were at play.
5. '[E]valuer l'état de conservation des habitats et identifier les activités humaines susceptibles de garantir leur pérennité ou, à l'inverse, de leur porter atteinte' (DIREN 2003, p. 3).
6. The reasons for the delay suffered in all three sites would take too much space to deal with here. Local environmental groups blame the inability of the préfet to reach agreement on the demarcations and management of the sites (Interview 13 April 2006).
7. 'Ce choix va à l'encontre du projet NATURA 2000 et de la preservation du parc *Marais Poitevin*' (Martin 2005b, p. 5).
8. 'Alain Mathieu contribue au défigurement du Marais Poitevin' (Riand 2005).

9. Royal initially received criticism from associations and the media for not involving herself at an earlier stage as the president of the Poitou-Charentes region (Biard 2005b). A press communication from the association CIMES declared that Royal should be awarded a 'golden star for deception' ('palme d'or de la deception') as they suspected her involvement in initially promoting the incinerator (CIMES 2005).

10. '[J]' exprime mon étonnement d'installer ce type d'équipement sur le territoire du parc ... [c]' est compliquer de demander des efforts aux agriculeurs dans le cadre de Natura 2000 et de voir arriver un incinérateur' (Girard 2005).

11. 'We are particularly worried about the proposed development of an incineration plant ... rejection of the waste incineration option is imperative' (CPAI 2005).

References

ADEME (Agence de l'Environnement et de la Maîtrise de l'Energie), 2008. Contrats Territorial Déchets: Syndicat Mixte du Pays Thouarsais, Poitiers: Délégation Régionale Poitou-Charentes.

Appleton, A., 2000. The new social movement phenomenon: placing France in comparative perspective. *In*: R. Elgie, ed. *The changing French political system*. London: Frank Cass, 57–75.

Baroux, P., 2005. L'Autre 'Révolution Orange': les opposants à l'incinérateur ont présenté hier les alternatives au projet. *Sud Ouest Dimanche,* 20 March, Environnement, pp. 8–9.

Bechmann, R., 2002. Le mouvement 'Aménagement et nature' dans la naissance et l'évolution de la politique de l'environnement, en France, dans les années 1960–2000. *Géographie, Economie, Société*, 4 (4), 355–361.

Bell, L., 2001. Interpreting collective action: methodology and ideology in the analysis of social movements in France. *Modern and Contemporary France*, 9 (2), 183–196.

Beroud, S., Mouriaux, R., and Vakaloulis, M., 1998. *Le mouvement social en France*. Paris: La Dispute/Snedit.

Bertolini, G., 1998. La politique française des déchets'. *In*: B. Barraqué and J. Theys, eds. *Les Politiques d'Environnement: Evaluation de la première generation 1971–1995*. Paris: Editions Recherches, 171–188.

Biard, S., 2005a. Action contre 'incinérateur: Coulon rebaptisé 'Dioxine City'. *La Nouvelle République du Centre-Ouest*, 31 January, Société, p. 8.

Biard, S., 2005b. Incinérateur: la confrontation? *La Nouvelle République du Centre-Ouest*, 3 January, Vie de la Cité, p. 5.

Boyer, J., 2006. Le Choix des Armes. *Sud Ouest,* 25 February. Politique, pp. 3–4.

Brousse, J., 2005. *Incinération des déchets ménagers: La grande peur*. Paris: Le Cherche Midi.

Clark, B., 2002. The indigenous environmental movement in the United States. *Organization and Environment*, 15 (4), 410–444.

Coban, A., 2004. Community-based ecological resistance: the Bergama movement in Turkey. *Environmental Politics*, 13 (2), 438–460.

CCIIPAC, 2003. *Lempdes (63): Un 'Non' Massif à l'incinérateur*. Press communication, 19 June. Clermont-Ferrand: ADEL.

CIMES, 2005. *L'incinérateur à Coulon: Cest à qui la faute?* 21 February. Melle: Citoyens pour l'Information dans le Mellois sur l'Environnement et la Santé.

CPAI, 2005. *Lettre Collective de 124 Médcins du Secteur de la CAN*. 27 January. Niort: Collectif pour l'Alternative de l'Incinération.

Davies, A., 2005. Incineration politics and the governance of waste. *Environment and Planning C: Government and Policy*, 23, 375–398.

Davies, A., 2007. A wasted opportunity: civil society and waste management in Ireland. *Environmental Politics*, 16 (1), 52–72.

Davies, A., 2008. Civil society activism and waste management in Ireland: the Carranstown anti-incineration campaign. *Land Use Policy*, 25 (2), 161–172.

DIREN, 2003. *Document d'Objectifs Natura 2000 du Marais Poitevin*. Vienne: La Préfecture de la Région Poitou-Charentes.

DGEMP, 2005. *La Valorisation énergétique produite par l'incinération des ordures ménagères*. Paris: Ministère de l'Economie, des Finances et de l'Industrie.

Dryzek, J., *et al.*, 2003. *Green states and social movements: environmentalism in the United States, United Kingdom, Germany and Norway*. Oxford University Press.

Duriez, H., 2004. Modèles d'engagement et logiques de structuration des reseaux locaux de la gauche mouvementiste à Lille. *Politix: Revue des Sciences Sociales du Politique*, 17, 165–199.

Edwards, B. and McCarthy, J., 2004. Resources and social movement mobilization. *In*: D. Snow, S. Soule, and H. Kiesi, eds. *The Blackwell companion to social movements*. Oxford: Blackwell Publishing, 116–152.

European Commission, 2005a. *N2000 zones for special protection in France: Chaine des Puys FR8301052*. Luxembourg: Office of Publication for the European Communities.

European Commission, 2005b. *N2000 zones for special protection in France: Le Marais Poitevin FR5410100*. Luxembourg: Office of Publication for the European Communities.

European Commission, 2006a. *N2000 zones for special protection in France: Pays des Couzes FR8312011*. Luxembourg: Office of Publication for the European Communities.

European Commission, 2006b. *N2000 zones for special protection in France: Vallées et côteaux thermophiles au nord de Clermont-Ferrand FR8301036*. Luxembourg: Office of Publication for the European Communities.

Eurostat, 2006. *Municipal waste by type of treatment: Metadata SDSS*, TSIEN130, Luxembourg: Office for Official Publications for the European Communities.

Franquemagne, G., 2007. From Larzac to the altermondialist mobilisation: space in environmental movements. *Environmental Politics*, 16 (5), 826–843.

Freeman, J., 1979. Resource mobilization and strategy: a model for analyzing social movement organization actions. *In*: M. Zald and J. McCarthy, eds. *The dynamics of social movements: resource mobilization, social control, and tactics*. Cambridge: Winthrop Publishers, 167–190.

Gerrard, M., 1996. *Whose backyard, whose: risk: fear and fairness in toxic and nuclear waste siting*. Cambridge, MA: MIT Press.

Girard, P., 2005. Incinérateur, Natura 2000 et label. *La Nouvelle République du Centre-Ouest*, 23 February, Environnement, p. 7.

Giugni, M., 2001. l'impact des mouvements ecologiste, antinucleaire et pacifiste sur les politiques publiques. *Revue Francaise de Sociologie*, 42 (4), 641–669.

Hayes, G., 2006. Vulnerability and disobedience: new repertoires in French environmental protests. *Environmental Politics*, 15 (5), 821–838.

Hayes, G., 2002. *Environmental protest and the state in France*. Basingstoke: Palgrave.

Hunter, S. and Leyden, K., 1995. Beyond NIMBY: explaining opposition to hazardous waste facilities. *Policy Studies Journal*, 23, 601–619.

INSEE, 2006. *Populations légales 2006 pour les régions et les départements*. Paris: L'Institut National de la Statistique et des Etudes Economiques.

ISWA, 2001. *Energy from waste statistics, state of the art report*. 1st ed. Paris: International Solid Waste Association.

ISWA, 2007. *Energy from waste statistics, state of the art report*. 6th ed. Paris: International Solid Waste Association.

Jacques, M., 2003. Non à l'incinération à gueugnon: pas de trêve pour les opposants au project d'implantation de l'incinérateur. *Hérault Tribune*, 13 May, Environnement, p. 8.

Kloek, W. and Jordan, K., 2005. *Waste generated and treated in Europe: data 1995–2003*. Edition 05. Luxembourg: Office for Official Publications of the European Communities.

Kriesi, H., 2004. Political context and opportunity. *In*: D. Snow, S. Soule, and H. Kiesi, eds. *The Blackwell companion to social movements*. Oxford: Blackwell Publishing, 67–90.

Kubal, T., 1998. The presentation of political self: cultural resonance and the construction of collective action frames. *Sociological Quarterly*, 39 (4), 539–554.

Le Grand, J., 2004. *Pour une mise en valeur concertée du territoire Natura 2000*. Vol. 1. Paris: Ministère de l'écologie et du développement durable.

Lefèvre, S., 2005. Alain Mathieu reste serein: le souhait du Parc du Marais de voir abandonner le projet d'incinérateur. *Sud Ouest*, 28 February, Environnement, pp. 5–6.

Leonard, L., 2006. *Green Nation: The Irish Environmental Movement from Carnsore Point to the Rossport Five*. Drogheda: Greenhouse Press.

Lepellier, L., 2008. *La démarche territorial: elements de la contractualisation et le Contrat Territorial Déchets (CTD)*. Paris: ADEME.

Lyvinec, G., 2004. Les Déchets seront brûlés. *La Nouvelle République du Centre-Ouest*, 10 December, Déchets, p. 2.

Martin, C., 2005a. Le collectif d'opposition à l'incinérateur réclame un débat contradictoire. *La Nouvelle République du Centre-Ouest*, 22 January, Déchets, p. 8.

Martin, J., 2005b. Ce choix est une provocation vis-à-vis de la vendée et des populations de Benet et des environs. *La Nouvelle République du Centre-Ouest*, 9 February, Société, p. 7.

Martin, V. and Olivier, G., 2005. Le compostage à vias: une des solution à la problématique de nos déchets. *Hérault Tribune*, 29 October, Environnement, pp. 6–7.

Mathieu, J., 1992. *La défense de l'environnement en France*. Paris: PUF.

McCauley, D., 2007. Environmental mobilisation and resource-opportunity usage. *French Politics*, 5 (4), 333–353.

McCauley, D., 2008. Governance and sustainable development: promoting the inclusion of civil society. *European Environment*, 18 (3), 152–167.

MEDD, 2008a. *Déchets: carte des incinerateurs en fonctionnement*. Paris: Ministere de l'Environnement et de Developpement Durable.

MEDD, 2008b. *Déchets: carte des incinerateurs en projet*. Paris: Ministere de l'Environnement et de Developpement Durable.

Moinet, D., 2007. *Déchets ménagers: Guide du compostage et la carte sites de traitement*. 49. Niort: Communaté d'Agglomération de Niort.

Morel, C., 2005. Label Perdu: l'incinérateur sur le territoire du parc. *Sud Ouest*, 11 January, Environnement, pp. 6–7.

Riand, E., 2005. Les opposants à Alain Mathieu lui souhaitent la fête des maires. *La Nouvelle République du Centre-Ouest*, 30 May, Niort, pp. 7–8.

Robert, C., 2006. Manifestation contre le futur incinérateur clementois. *Libération*, 13 March, Terre, p. 11.

Rootes, C., 2006. Explaining the outcomes of campaigns against waste incinerators in England: community, ecology, political opportunities, and policy contexts. *Research in Urban Policy*, 10, 179–198.

Saarikoski, H., 2006. When frames conflict: policy dialogue on waste. *Environment and Planning C: Government and Policy*, 24 (4), 615–630.

Sauges, O., 2006. *La naissance de P.R.O.P.R.E.: collectif pour repenser l'organisation et le plan de recyclage et d'elimination des déchets*. Clermont-Ferrand: Député du Puy-de-Dôme.

Szarka, J., 2002. *The shaping of environmental policy in France*. New York: Berghahn.
Tarrow, S., 1998. *Power in movement: social movements and contentious politics*. Cambridge University Press.
Van der Heijden, H.-A., 1997. Political opportunity structure and the institutionalisation of the environmental movement. *Environmental Politics*, 6 (4), 25–52.
Visiglio, G.R. and Whitelaw, D.M., eds., 2003. *Our Backyard: a quest for environmental justice*. Lanham, MD and Oxford: Rowman and Littlefield.
Walsh, E., 1988. New dimensions of social movements: the high level waste-siting controversy. *Sociological Forum*, 3 (4), 46–65.
Walsh, E., Warland, R., and Clayton-Smith, D., 1997. *Don't burn it here: grassroots challenges to trash incinerators*. University Park, PA: Penn State Press.
Waters, S., 2003. *Social movements in France: towards a new citizenship*. Basingstoke: Palgrave Macmillan.

Interviews

Interview with anonymous representative, *La Mairie de Bênet*, civil servant charged with waste management (13 February 2006 in La Roche-sur-Yon).
Interview with unnamed representative, *Association Contre l'Implantation de l'Incinérateur à Proximité de l'Agglomération Clermontoise*, policy officer on waste and transport infrastructure (13 April 2006 in Lyon).
Interview with unnamed representative, *Deux-Sèvres Nature Environnement*, policy officer on waste management issues (30 August 2006a in La Roche-sur-Yon).
Interview with Denis Gaborieau, *Confédération Paysanne de Vendée*, departmental representative (30 August 2006b in La Roche-sur-Yon).

Grassroots mobilisations against waste disposal sites in Greece

Iosif Botetzagias[a] and John Karamichas[b]

[a]Centre for Environmental Policy and Strategic Environmental Management, Department of Environment, University of the Aegean, Greece; [b]School of Sociology, Social Policy and Social Work, Queen's University, Belfast, Northern Ireland

The government of Greece has gained notoriety for its failure to implement EU environmental directives in general, and is criticised specifically for its lack of an effective plan for the safe disposal of waste. Local mobilisations against a series of planned 'Sanitary Waste Disposal Sites' (HETAs) in three municipalities of Attica are examined. Should such protests be classified as NIMBY (not in my backyard)? Or do they present broader claims of justice and equity? Qualitative analysis of the protesters' on-line campaign material reveals that while these mobilisations do demonstrate some NIMBY characteristics, such campaigns should rather be perceived as *ad hoc* mobilisations reflective of tensions of late modernity. The public's mistrust of science and concerns about democratic deficit and accountability, as well as different perceptions of risk, are prominent.

Introduction

According to Kasperson (2005, p. 13), 'few problems have proven more contentious or perplexing than the siting of hazardous facilities'. What is truly remarkable is that public concern, challenges and opposition to the siting of hazardous facilities is shared across different national contexts. As expected, the intensity of this experience is very much conditioned by the policy-making styles adopted by the administration, including open or closed systems (see Kitschelt 1986), the support of international Environmental Non-governmental Organisations (ENGO) and political forces for the challengers, among other variables. In almost all cases the need for the facility in question seems clear to its advocates and supporters but this sentiment is rarely shared by the residents of the host communities due to, among other things: inequities in the distribution of costs and benefits; risks and risk perception; lack of clarity about the need for such facilities; lack of trust in expert authorities (Wright 1993, Kasperson 2005).

Although similar mobilisations have occurred around Greece, here we study grassroots mobilisations against the siting of new, state-of-the-art landfills (or 'Sites for Sanitary Disposal of Waste', HETAs to use the acronym for the equivalent Greek words) in the Attica region, where the Greek capital city, Athens, is situated. One issue of importance when dealing with these mobili-sations is their location in the general context of campaigns against LULUs (locally unwanted/undesirable land use) across the world. Should we classify the protests against waste disposal sites in the Greater Athens area as NIMBY (not in my backyard) contestations with some or all of the characteristics usually attributed to them, or we can trace the development of something resembling or approximating the wider environmental justice movement?

At first glance, and since the proposed HETAs are to replace a much larger number of existing, illegal and notoriously noxious 'landfills', or, rather more appropriately, 'dumping sites', such local opposition can all too easily be seen as instances of NIMBYism. As social phenomena, NIMBY responses to LULUs have been variously characterised, both negatively and positively. Viewed negatively, such protests are seen as reflections of the narrow self-interest of the opponents of these facilities; viewed positively, they can be seen as a celebration of civic democracy (Schively 2007, p. 257; see also McClymont 2008 for a discussion of how close these two perceptions actually are). Yet, lately, the usefulness of the NIMBY concept as a research tool has come under increased criticism; it has been argued that 'while the language of NIMBYism is rife amongst participants in [local development] disputes, those interested in understanding, managing or resolving such conflict should abandon the term' (Burningham 2000, p. 55), and that it is more 'an empty concept' than 'an explanation' (Wolsink 2006, p. 85; see also Hubbard 2006 for a reply to Wolsink's critique).

We consider the complete rejection of the NIMBY concept unwarranted. This is because various studies have shown, time and again, that opposition to landfill siting does stem from reasons that can be considered to be of a NIMBY nature: concerns over physical dangers (such as pollution and threats to health), aesthetic concerns and a feeling of uneasiness at becoming the area's 'garbage dump', concern over falling property values as well as suspicion and low trust in decision makers, developers and operators (Bacot *et al.* 1994, p. 230, Cavatassi and Atkinson 2003, pp. 29–30). Such reasons are in accordance with the generic perspectives of NIMBYism (Burningham 2000, pp. 56–58) which view the phenomenon as 'an ignorant or irrational response', as 'a selfish response' or as a 'prudent' one.

Yet we also have serious reasons to question the concept's appropriateness. To start with, even the latter two of the NIMBY perspectives previously mentioned do not look so bad or irrational under close scrutiny. Is not selfish behaviour the alter ego of the rational actor behaviour in free-market systems?

And since when is prudence a bad thing (Burningham 2000, pp. 57–58)? Furthermore, we have cases where local opposition is articulated in terms than can hardly be labelled as 'irrational' or 'selfish', but is rather based on values, such as notions of equity, and on very rational calculations, such as the (perceived) need for the proposed development (Lober 1993, Lober and Green 1994, Rootes 2007). Also, as Rootes (2007, p. 733) rightly points out, local environmental campaigns, depending on the national context, develop not only along the 'environmental faultline' but also along other prominent societal ones (such as on issues of rights and democracy). Things are further complicated if we consider that NIMBY and other reasons for opposition are usually difficult to separate. This is either because they are interwoven (Cavatassi and Atkinson 2003) or because locals, in their *conscious* attempt not to be labelled/seen as 'NIMBYs' (Burningham 2000), mask their true reasons for concern with socially acceptable ones.[1]

However, there also exist broader issues pertaining to grassroots mobilisations which the NIMBY frame may miss altogether. As Freudenberg and Pastor (1992, p. 39) argue, research on local development disputes should move beyond NIMBY-related explanations and instead 'focus on understanding the broader system that creates such conflicts in the first place'. One line of research of particular relevance to our analysis of grassroots mobilisations against landfill siting is the role played by the decision-making process (see Burningham 2000, pp. 58–59), and the use therein of such notions as 'science' and 'expertise'.

We have already mentioned that in many local mobilisations, citizens have been particularly distrustful of developers and state officials alike. One study in the US (Bacot *et al.* 1994, p. 235) found that the percentage of citizens more inclined to favour a landfill siting dropped when the proposed authority responsible for inspections during operation moved from federal, to state, to county government, only to rocket to 75% 'if citizens themselves, through a local committee, are empowered to close the landfill if deemed unsafe'. It is obvious that citizens are more inclined to accept LULUs if they feel they have a meaningful say over the development, its assignment and management. Various studies have pointed out that locals are less likely to protest if they consider both the siting process and the distribution of costs to be fair and equitable (Lober 1993, 1995, Lober and Green 1994) or when they trust the government (Hunter and Leyden 1995). Yet the process can hardly be perceived as such when waste management plans are taken away from the elected local authorities, as has been the case in Greece (Andreou 2004, pp. 9–10). In a similar development in Ireland, such a move was perceived by opposition MPs as a sign that, whatever the explanations provided, the government 'simply wanted plans adopted irrespective of the local appropriateness of the strategies contained within them' (Davies 2007, p. 82). In such a context, citizens come to assume that they are dragged into an 'unfair' deal, or that they are being 'victimised', presumably due either to their community characteristics or to the lack of powerful allies who could shield them against the undesirable development.

The perceived or alleged 'unfairness' of waste facility sitings has, in the US, been highlighted by what has consolidated into the environmental justice movement (EJM). The EJM emerged in the United States in the 1980s and can be seen as a reinvigoration of the demands of the civil rights movement in a new political arena (see Capek 1993, Pellow 2000). The environmental justice frame highlights a different perspective on the opposition to LULUs as it acknowledges the disproportionate health and environmental impacts that are intentionally or unintentionally inflicted on working-class, low-income or ethnic minority communities by the siting of waste disposal facilities. Yet, besides the 'particularism' of opposing specific cases of acute environmental degradation and 'intended' environmental injustice, some scholars have argued that the EJM is also concerned with the injustice resulting from the 'systemic failings in legal, regulatory and economic regimes' (Watson and Bulkeley 2005, pp. 413–414). At a parallel level, it is debated to what extent the EJM is concerned with 'consequential questions' alone (issues of fair distribution of bads and goods) or whether it takes into account 'procedural questions' (pertaining to issues of decision making and community involvement).

Both NIMBY and environmental justice frameworks (especially in their broader, 'systemic' and 'procedural' form) can coalesce around the risk concerns put forward by the local protesters and their supporters and their evident lack of trust in the claims made by experts. Risk and the lack of trust surrounding expert authorities are the two central components of reflexive modernisation, a very influential contribution to social theory developed independently, albeit in parallel, by Ulrich Beck (1992, 1995) and Anthony Giddens (1990, 1991) to account for the changes and challenges of late modernity, with the environmental problematic constituting their main point of reference. For Beck, the relationship between scientific expertise and environmental protection is characterised by an evident ambivalence. On the one hand, science as applied technology is the principal issue behind many environmental hazards and risks facing contemporary societies. On the other, science possesses the necessary tools for both identifying the existence and calculating the possible consequences of these risks while it has the capacity to offer solutions for the amelioration of environmental problems. However, the uncertainty that characterises scientific hazard assessment is unavoidable, and this, for Beck, has been aptly demonstrated by some high-impact manufactured environmental disasters of the late twentieth century, such as Chernobyl, Bhopal and the BSE crisis. As a result, the deference to scientific reasoning that supposedly prevailed in earlier periods of modernity has been shaken. Giddens, along the same lines, discusses the process by which the normalcy desired by people is very much dependent on an uncomplicated trust in experts and their effective control of technological systems. Events, outside local control, break down this trust and lead to the creation of ad hoc local community groups.

Two works have seized the opportunity offered by reflexive modernisation in general and the *risk society* thesis in particular to account for the con-testation arising out of opposition to facility siting (Haris Ali 1997, Baxter *et al.*

1999), which, after all, is an event usually orchestrated outside local control. Both see how the risk associated with any particular case of facility siting can be framed as an issue of social equity with an overwhelming suspicion towards the (re)assurances offered by the experts involved. The former appears momentarily to challenge Beck's position on the equalising effect of environmental risks: 'poverty is hierarchic, smog is democratic' (Beck 1992, p. 36). Indeed, this widely cited position has attracted the great bulk of criticisms of the *risk society* thesis as it appears to disregard both the environmental justice movement and the environmentalism of the poor. With that in mind, it is accepted that Beck's position seems to be unsubstantiated when applied at the local/regional level and on a small time frame, but 'since risk is most often played out at the local/ regional level the concepts [mistrust of expert authorities/science] are worth exploring at this scale' (Baxter *at al.* 1999, p. 94).

Mistrust of expert authorities/science can be viewed as lack of trust toward the local/national political and technical decision makers, or as a generalised mistrust 'in the system of science and technology (to detect and/or monitor the risks)' (Harris Ali 1997, p. 483). There is an assumption underpinning the reflexive modernisation perspective in general (and the risk society thesis in particular) that in early modernity trust in expert systems prevailed. If we follow this line of reasoning, then grassroots opposition against HETAs can be seen as one local manifestation of the aforementioned general mistrust that one can detect in late modernity. Nevertheless, as Wynne (1996) argues, rather than accepting the lack of open contestation of expertise in the past as evidence of trust, as, in a reverberation of Almond and Verba's (1963) *Civic Culture*, both Beck and Giddens appear to do, it is much more sensible to view it as indicative of an opportunity structure lacking sufficient access points by which grassroots actors might challenge the experts. Moreover, in the light of recent discussions of the 'Mediterranean Syndrome' (La Spina and Sciortino 1993) and the 'Southern Problem' (Pridham and Cini 1994), is it possible that mobilisations against HETAs are manifestations of ad hoc mobilisations characteristic of late modernity, or do they instead conform to some typically southern and/or Greek characteristics of essentially NIMBY environmental protests?

In the next two sections we present the framework concerning municipal management in Greece and we offer an overview of the debate concerning the mobilisations against landfill siting in Attica. In the third section we focus on the mobilisations, offering both a quantitative analysis based on newspapers' reporting of protest events, and a qualitative analysis based on interviews, press releases, content analysis of web pages of groups involved in the protest. In the concluding section we summarise our findings and discuss the appropriateness of labelling these mobilisations as 'mere NIMBYism'.

Municipal waste disposal in the Greek context

The issues of waste production and disposal in Greece are increasingly urgent. In line with the axiom stating 'the richer you are, the more garbage you produce', waste production in Greece demonstrated a sustained rise over the last two decades (EEA 2008) (see Figure 1). This increasing amount of waste has been landfilled, while in many cases it was simply dumped at illegal and uncontrolled sites. As late as 2005 Greece was landfilling almost 90% of the waste produced (EEA 2007, p. 288). The transposition of the Landfill Directive (99/31/EC) occurred, after a delay, in late 2002, when infringement procedures against Greece were already under way, and it was, as has been the usual Greek practice, 'reproduced "word-for-word", and approved by all stakeholders without much debate' (Andreou 2004, pp. 5–6).

Waste reduction and recycling still lag behind the EU average, although some positive developments have occurred over recent years. Following the transposition of EU Directive 94/62/EC 'On packaging and packaging waste' by Law 2939/2001, Greece established the 'National Organisation for the Alternative Management of Packaging and Other Waste' and two years later, by Ministerial Decision 106453/20-02-2003, approved a nationwide scheme for managing packaging in particular and urban waste in general, coordinated by the Hellenic Recovery Recycling Corporation (HERRCO), a joint venture of the Central Union of Municipalities and Communities of Greece (KEDKE) and some of the largest Greek companies. Nevertheless, reduction at source was 'not set as a policy objective', the main concern being 'the development of controlled elimination schemes' (Buclet and Godard 2001, p. 306). Yet HERRCO has made remarkable progress: in autumn 2007, some 1200 enterprises were registered with it while the number of cooperating local authorities tripled (HERRCO 2007).

However, the majority of urban waste has ended up in uncontrolled landfills, which pose a health and environmental hazard. Greece made EU legal history in July 2000, when it became the first EU member state to fail to

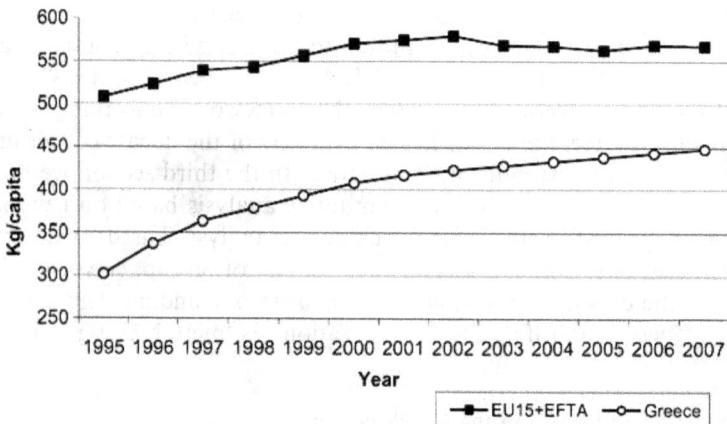

Figure 1. Municipal waste generation in western European countries (EU 15 + EFTA) and Greece (kg/capita). *Source*: EEA (2008).

comply with a ruling of the European Court of Justice (ECJ) and was fined €20,000 per day for violations of the Waste Framework Directive (75/442/EEC, as amended by Directive 91/156/EEC) in respect of the Kouroupitos landfill site, near Chania, Crete.[2] The Kouroupitos case speeded up the hitherto crawling governmental attempts to shut down illegal landfills and to replace them with state-of-the-art HETAs. By 2004 an estimated 1453 uncontrolled landfills were still operating, with a further 1173 closed and awaiting rehabilitation (YPEHODE 2005, p. 3). A phase-out plan was set, aiming to close and rehabilitate all of them by the end of 2008, at a total cost of around €0.5 million (YPEHODE 2005, p. 14). This plan was set in the face of yet another ECJ ruling (Case C-502/03), pointing to Greece's various failings regarding the Waste Framework Directive. At the time of writing it is not certain whether Greece will manage to meet the end 2008 deadline; state officials have recently stated that in the best case scenario around 500 landfills will fail to meet it, thus incurring a penalty of €34,000 per day per landfill (Hatziioannidou 2008).

The present crisis point has been the result of poor planning and inadequate funding by the state agencies (Agapitidis and Frantzis 1998, p. 246, Andreou 2004), the institutional fragmentation of urban waste management,[3] and determined local opposition to the development of new, properly planned and managed landfills. The situation in the Attica region, which we will proceed to analyse in detail, is a case in point.

The debate about HETAs siting in the Greater Athens region

The first organised waste landfills for the Attica region were the ones at Schisto (opened in 1960) and Liossia (opened in 1965) in western Attica, one of the most underdeveloped areas of the country; the value of land is the lowest in all Attica, while the unemployment rate is one of the highest, in an area inhabited mainly by blue collar workers and 'marginal socio-economic categories' (i.e. land labourers, itinerant salesmen, etc.) (Maloutas 2000). In line with international experience, for over four decades in Attica the 'flow of waste, [ran] downhill along the power gradients of society' (Ackerman and Mirza 2001, p. 114). Nevertheless, this flow of waste was also an 'asset': in 1996, the area was receiving approximately €0.5 million (at current prices) per annum for taking the municipality of Athens' urban waste (Eleftherotypia 1996). Since 2005, a community wishing to dispose of its waste in the Liossia landfill has to pay a flat fee, amounting to 4.5% of the community's annual revenue. This fee is collected, and the revenue advanced to Liossia's municipal authority, through ESDKNA (Association of Communities and Municipalities in the Attica Region) (Elafros 2005).

ESDKNA[4] was founded in 1970 with the main aim of managing Attica's municipal waste, and in the early 1980s tried to find new landfill sites, while suggesting – for the first time – alternative waste disposal methods, such as composting and incineration (with energy recovery). Yet no measures were taken for over 20 years. When in 1991 the Schisto landfill finally closed, six years after reaching its capacity, Liossia was left as the only landfill receiving almost all of Attica's municipal waste. Although over the next decade there had been suggestions that Attica's waste should be exported and possibly incinerated at a plant in the adjacent region of Viotia, both Ministers for the Environment over the period had ruled out these options: as one of them, Kostas Laliotis, bluntly put it as early as 1993: 'one prefecture should not export its waste outside its boundaries', while incineration was ruled out as 'posing a great risk to human health'.

A number of studies, local mobilisations and appeals to the administrative courts occurred during the 1990s, yet no real progress was made. A major turning-point was reached in 2003. The new Minister for the Environment, Vasso Papandreou, made it clear that if the local government bodies did not manage to decide on the issue themselves, she was willing to pass a law ending the stalemate. Despite the usual mobilisations, in June 2003 the Greek Parliament, following a high-spirited debate, by a narrow margin approved six areas which should host the three new HETAs for Attica. Three of them were duly selected over the summer of 2003, the remaining three being labelled as 'reserves'.

The ECJ's ruling of autumn 2005, and the looming fine, added an element of urgency to the debate. The affected Attica communities appealed once again to the Council of State (the Greek Supreme administrative court), only for the Court to reject their appeals in spring 2007 (Dalamagka 2008). In July 2008 one of the new HETAs became operational, at Fili (western Attica), at a site adjacent to the existing Liossia landfill. At the opening ceremony it was mentioned that if the remaining two HETAs – in northern and eastern Attica – were not established, the existing one would reach its designed capacity in just three years, that is, six years earlier than planned (Selamazidis 2008). At the time of writing the works on these two sites had not yet commenced, while the local inhabitants were preparing to take direct action to fight back any attempt to establish a construction site, even against the Greek Police Riot Squads (MAT). The three-month stand-off between local protesters and the riot squads on the island of Corfu over a similar development – the Lefkimi HETAs – with violent clashes, a number of citizens and policemen injured, and a mother of two accidentally killed (in May 2008), serves as a reminder of how such a confrontation can all too easily get out of hand. Table 1 presents the developments over the last 40 years in brief:

NIMBY or beyond? An analysis of the protest against the HETAs siting in Athens

The quantitative data for the Attica mobilisations comes from an analysis of protest cases as reported in the Greek daily *Eleftherotypia*, a centre-left newspaper, over the last 17 years. We have used the on-line version of the newspaper and conducted our data mining by means of a keyword search.

Table 1. The debate around municipal waste landfills in Greater Athens Region.

1960	Schisto landfill begins to operate (western Attica)
1965	Liossia landfill begins to operate (western Attica)
1970	ESDKNA (a regional waste management body) is created
1980s	First attempts to secure new landfill sites
1991	Schisto landfill closes
	ESDKNA assigns the first studies for Avlonas and Grammatiko (northern Attica) to a private firm
1992	The Ministry for the Environment includes three more sites: Varnavas (N. Attica) and two at Ritsona, outside Attica's boundaries, in Viotia.
	TEDKNA (Attica's Local Municipalities Federation) agrees to establish three HETAs – one of them at Liossia
1993	Ritsona is ruled out by the new Minister, as well as any alternative waste disposal techniques (e.g. incineration)
1996	*Jan.*: ESDKNA assigns new studies for nine sites to the Polytechnic University of Athens
	Spring: Liossia landfill starts its periodic closures as a pressure tactic
	Summer: The Ministry assigns new studies to the University of Thrace
	Dec.: The Thrace team announces that no site is suitable
1997	*Jan.*: Attica's Regional Council pre-selects three sites in north-eastern and south-eastern Attica
	May: The Ministry approves the relevant Environmental Impact Studies
	July: The Supreme Court accepts the locals' appeals and asks for a temporary halt to the projects
1998	*Dec.*: The plenary of the Supreme Court finally rejects the appeals on judicial grounds
2000	The Prefecture of Eastern Attica is allowed to proceed with its 'local' waste management plan
2001	*Sept.*: Attica's Regional Council re-introduces Eastern Attica to the 'Attica Region Waste Management' plan
2003	*Jan.*: The Minister of the Environment makes it clear she will solve the issue by law
	Spring: Major grassroots mobilisations
	June: The Greek parliament approves three new HETAs sites, one for each of western, north-eastern and south-eastern Attica (Liossia, Grammatiko and Keratea respectively)
	July: The affected communes appeal to the Supreme Court
2005	*May*: National Plan to phase out all landfills by the end of 2008
	October: The European Court of Justice condemns Greece for poor application of waste-related Directives
2007	*Spring*: The Supreme Court rejects the appeals of the north and eastern Attica communes
2008	*July*: The Liossia HETAs (the first of the planned three) opens

Using newspaper data to collect information on (past) mobilisation has been a common practice during recent years. As an example, we refer to the research conducted by an international team of scholars coordinated by Chris Rootes, which has used this method for studying the transformation of environmental activism in Western Europe (Rootes 2003). For the study of the Greek case, this same daily was used, since this newspaper is considered to adequately cover not only environmental but also domestic news (Kousis 2007). The availability of such research findings also allows us to place the local mobilisations against waste within the broader framework of Greek grassroots mobilisations. Thus, as Kousis demonstrates (2007, p. 796), domestic waste-related protest has been the largest sub-category of local protest cases (approximately a quarter of all cases) for the period 1974–1994, as well as the one attracting the highest political attention, as shown by the presence of political party representatives. Since, as we have already mentioned, the peak years had been 1996 and 2003, we could argue that the prominence of waste-related mobilisations vis-à-vis other reasons for protest has remained high throughout the period under investigation.

We do not here go into great detail concerning the analysis of these protest cases (e.g. claims, alliances, mobilisers' characteristics and so on) because this is an issue partly addressed elsewhere (Botetzagias 2006, Dalamagka 2008). For our analysis it is sufficient to present the mobilisers' claims, as recorded in the newspaper's coverage of relevant protest 'events' (PEs), ranging from 'conventional' PEs (e.g. public meetings), to 'confrontational' PEs (e.g. activities such as blockades/closure) and finally 'violent' events. Up to three claims were recorded for each 'event', which proved sufficient for all cases, and these are reported in Figures 2, 3, 4 and 5.

A superficial reading of the graph in Figure 2 would lead to the conclusion that we are dealing with NIMBY mobilisations. After all, the majority of claims raised in protests are directed against the proposed HETAs. Nevertheless, this general graphical representation masks the existence of two different campaigns, occurring in parallel in western and north-eastern Attica, each having different goals, employing different tactics and articulating distinct

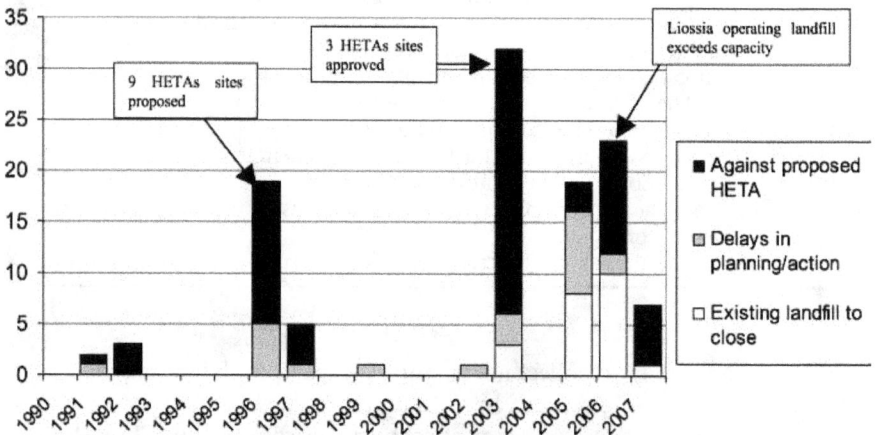

Figure 2.   Protest event claims for Attica's waste mobilisations and major waste-related events (1990–2007). *Source*: Newspaper *Eleftherotypia*, own coding (up to three claims per PE were coded).

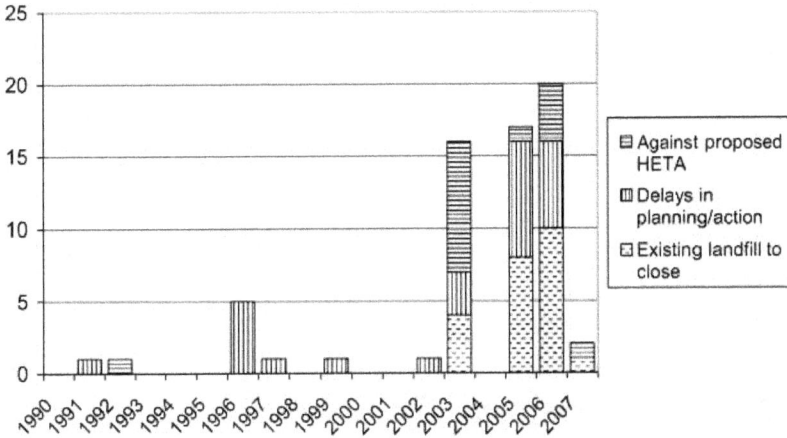

Figure 3. Protest events claims for western Attica's waste mobilisations (1990–2007). *Source*: Newspaper *Eleftherotypia*, own coding (up to three claims per PE were coded).

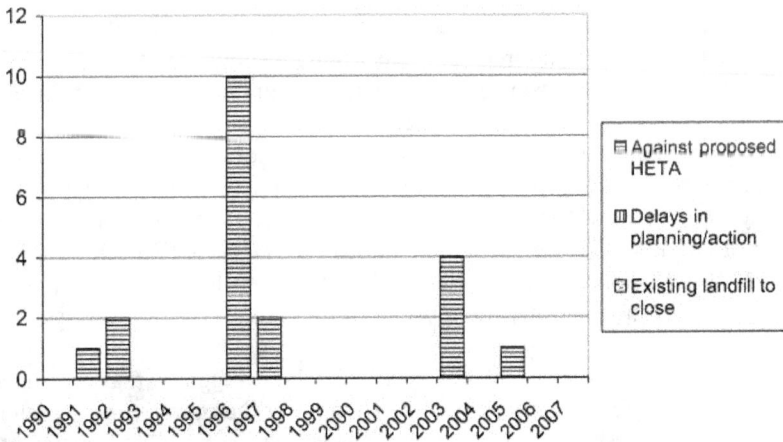

Figure 4. Protest events claims for northern Attica's waste mobilisations (1990–2007). *Source*: Newspaper *Eleftherotypia*, own coding (up to three claims per PE were coded).

discourses concerning waste management. In northern and eastern Attica, the locals' sole claim was 'against the proposed HETAs' (Figures 3 and 4). Yet in western Attica things were more complicated. The area has hosted Attica's only 'legal' landfills and, in the early 1990s, when it became obvious that new sites had to be assigned, the local municipality was offered a deal: on the one hand replacing the existing landfill – already then a notorious dumping site – with two new HETAs, thus securing the continuation of the generous financial subsidies; on the other hand, the promise to create a number of new HETAs in northern and eastern Attica, accommodating these areas' local needs and

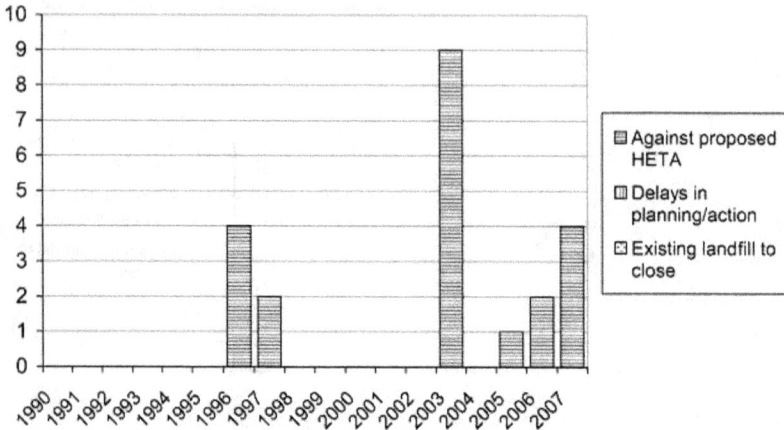

Figure 5.   Protest events claims for eastern Attica's waste mobilisations (1990–2007). *Source*: Newspaper *Eleftherotypia*, own coding (up to three claims per PE were coded).

relieving the burden on western Attica. The deal was accepted, but – as we have discussed earlier – not honoured. Thus, as Figure 5 shows, the local opposition, orchestrated by local government officials, mounted, along with protests against the delays in planning. However, the area also witnessed a parallel campaign of local inhabitants denouncing the local officials' support of government plans. This 'counter' protest campaign, which was supported at times by the western Attica prefecture, advocated a dual goal, demanding the closure of the landfill *as well as* not establishing HETAs (Botetzagias 2006). It made its first appearance in 2003, when the new HETAs were assigned, and from then on it became the most prominent campaign in western Attica (2005–2007).

The qualitative analysis of the available data suggests that in all sub-regional protests NIMBY echoes can be found, yet the mobilisers also refer to much broader issues, with northern and eastern Attica mobilisers articulating the most diversified discourse, bringing up issues of mistrust, scientific objec-tivity, environmental, health and aesthetic issues as well as equity. For eastern Attica, issues of justice and equity predominate.

Thus the opening page of the website 'No to HETAs' (http://www.ox-istoxyta.gr/, accessed 28 July 2008) of the Grammatiko community (northern Attica), argues that 'the existing national waste management plan, being simply the synthesis of partial Regional Waste Management Plans, is *scientifically unsound*, thus causing *major environmental and social problems* … as well as placing a *huge burden on the national economy* and the citizens' (emphasis original). However, it is also claimed that they would welcome, and bear the burden assigned to them by, 'a revised and demonstrably rational National Management Plan', based on the 'scientific/University studies', which 'suggest/ prove' that 'on any account, the cheapest and best solutions [lie] *outside Attica*'.

Thus some of the studies available through the website do argue that the selected area is not appropriate (Kelepertzis 2003) while others propose a revised national management scheme, suggesting the creation of larger waste treatment facilities throughout the country (only 10 instead of the 100 proposed HETAs) (Economopoulos 2007). It is also argued that the proposed management techniques are out of date and that the state should have been placing more emphasis on preventing waste generation and on recycling solutions.

In the relevant press releases, northern Attica's local government officials – spearheading the protest – have tried time and again to rationalise their objections and to disperse any ideas of NIMBYism:

> I know and understand that the waste issue has turned into a saga ... that the garbage should go to some place. ... We all want a solution. Yet we have to be careful. *None of us can accept a decision that is not a solution but a disaster* ... WE DO NOT SAY 'NO' JUST FOR THE SAKE OF IT. (Koukis, president of the Grammatikos' community, 2007a, emphasis original)

In similar vein, in another relevant blog (Oust the HETAs), the local citizens' committee of another northern Attica community were infuriated by what they perceived to be a biased representation of their struggle by a Greek TV 'Eco-news' programme. Their initial joy at being contacted to present their case was turned into dismay a few days later:

To Eco-News editorial team:

> We watched your reporting and we are really sorry since instead of bringing out Grammatiko's HETAs problem, you buried it! We feel almost deceived. You showed spokespersons from Grammatiko and Varnavas [a nearby community] having as their sole argument a blank refusal to the siting, without showing the substantial arguments that were mentioned and recorded, such as 'Why there should not be HETAs at Grammatiko' and 'What are the alternatives'. The comment of the representative of ESPA [Environmental Engineers Consultants, a private firm] was the final blow, presenting those reacting as being uninformed and having no relevant knowledge of the issue. (Active Citizens of Varnavas, http://energipolitesvarnava.blogspot.com/2008/07/tv.html, accessed 28 July 2008)

Issues of equity and fairness have also featured prominently:

> Mr Karamanlis [the Greek Prime Minister] my name is Nikos Kala ... and I live at Grammatiko ... please give some thought to my health and my future. Bringing the garbage to my village will pollute the area in which I live, the sea where I go in summer and you will ruin my health. Please give it some good thought. (excerpt from a letter sent by Grammatiko's schoolchildren to the Greek Premier; Koukis 2007b)

> On which ethical grounds does the Mayor of Athens, a city producing 30% of Attica's waste, demand to send his produce not simply to another municipality but to another Prefecture? (Koukis, president of the Grammatikos' community, 2007b)

Time and again the mobilisers refer to what they perceive as the 'imbalance' or the 'inequity' of having to host HETAs able to receive at least 900 tonnes per day when northern Attica's waste production is a mere 60 tonnes/day. Yet we should note that the Grammatiko community's ideas concerning equity issues are rather contradictory. This is clearly seen concerning the HETAs' siting. Although it is claimed that scientific studies suggest that all waste should be moved out of Attica, in these very same studies western Attica, the long-time garbage bin of Athens, figures prominently as the host area for one of these mega-plants. Furthermore, they never seem to wonder whether or why the export of their area's waste, although 'scientifically sound', is going to be accepted by the recipient areas' citizens – or how such an approach comes to terms with their insistence on 'equity'.

At the same time, environmental and aesthetic reasons are also stressed:

> If we let the landfill be constructed, we run the risk of polluting, due to the HETAs' toxic effluent, the whole of Evoikos gulf as well as Marathonas' Lake [Athens major water reservoir] as well as the area's groundwater.
>
> The access to the HETAs is only available through Grammatiko's narrow, winding roads; could you cope driving for hours behind the hundreds of waste-carrying trucks? (flies, filth, garbage!!!)[5]
>
> This area is the last green area of Attica. The northerly breezes, blowing for 240 days per year, now refresh and bring fresh air to the whole Athenian plateau. If the HETAs goes ahead, they will bring airborne carcinogenic pollutants. (Apostolopoulos 2008)

Similar views and rationalisations are to be found in the blogs/webpages voicing eastern Attica's opposition (Anti-HETAs 2008). Yet an issue featuring prominently in this blog, and setting it apart from the rest, is that of 'past/accumulated knowledge', spurred by a deep distrust of developers and state planners: citizens are wary of these new HETAs proposals because they seriously doubt that they will be any different from the existing landfills.

> The citizens of Keratea [eastern Attica] have struggled for 12 years against a HETAs' siting in their area, not only because it is an illegal development which does not meet the siting requirements, but because they know what a HETA in Greece is and how it operates in reality. Already, for the 'state-of-the-art' HETAs at Liossia, opened only last week ... there are complaints filed with the EU, concerning its faulty operation.[6] (Anti-HETAs 2008)

But again, these claims are deeply interwoven with amateurish attempts to take the argument further, and by taking it one step too far, in effect aiming to prejudice the reader. Thus, the blog features photos aiming 'to show you what a HETAs is, not in theory but in practice, through images of those already in operation'. Yet the photos provided are years old – and are not of HETAs but of the very landfills the HETAs are supposed to replace. Though the mobilisers have good and valid reasons to question the HETAs' proper functioning in the circumstances of Greece, they also try to construct an 'appalling' image of these developments, trying to relate them – at any cost – to a familiar, and notorious, reality. In a direct challenge of the developers' and state officials' claims of the 'newness' and 'safety' of the projects, the mobilisers try to equate HETAs with the old landfills.

> And yet this place, where you and your children can take a deep breath away from the cement of Athens, is at risk of an ecological disaster; because of the *new landfill (they now call it HETAs)*. (Apostolopoulos 2008, emphasis added)

For their part, state officials are trying to sugar-coat the siting and operation of the schemes. Arguably losing the 'framing' war over the new waste disposal sites,[7] and facing an EU legislation which no longer supports the construction of HETAs, the cabinet decided that the selected sites in eastern and northern Attica will operate as HETAs until the necessary infrastructure (e.g. recycling centres, sorting plants etc.) is built nearby. From then on, the sites will be turned into HETE (Sites for Sanitary Disposal of Waste Remnants), receiving only what remains after the recyclables and compostables are removed (*Eleftherotypia*, 23 July 2008). The junior Minister for the Interior stressed in 2008 that 'No-one is talking anymore of [establishing] HETAs', while it is exactly for such kinds of projects that all relevant permits and planning have been approved, a fact quickly picked up by the mobilisers and ridiculed in a number of YouTube videos (Kollias 2008).

Finally, the western Attica campaign is the least visible: we find no citizens' blogs,[8] no assigned studies contradicting those of the developers, no YouTube videos. Since for that area the HETAs was broadly portrayed by developers as an 'improvement', decommissioning the existing notorious landfill in favour of a 'sanitised' HETAs, there was far less room for arguing that this was a 'deterioration' – as mobilisers in eastern and northern Attica did. Furthermore, all scientific studies of the past 30 years did allow for a waste treatment facility in western Attica, in the very same areas that have been used for decades, thus denying scientific credence to any opposition. Thus, the only course open to local objectors was to refer to issues such as their area's overall environmental and social degradation.

> You are the 'Reportage without Frontiers'; we have lived in the Landfill without Frontiers. For 45 years, can you imagine that? Living under all Attica's trash. We do not dare to open our windows. Our children cannot enjoy what every other child can. (St. Kolliou, Local women's committee for removing the landfill, 2008)

> Everyone should take his responsibilities, carry his fair share of the burden and be responsible for what he produces ... The mentality that western Attica is a second-class area and its citizens have fewer rights than the rest should go ... We are determined to fight to stop this injustice against our area. There should finally be an end to the disregard and systematic abasement of our area. Western Attica can no longer be the area where all of Attica's LULUs are assigned, over and over again. (Arkoudaris, Prefect of Western Attica, 2008,)

> I hear empty promises. We have made fools of ourselves, discussing just for the sake of it. There are no solutions for those who do not want to find any. Enough! The garbage must go now. We are not Attica's castaway children ... We [Liossia's municipality] will find solutions for our own waste ... [Other] Mayors should do the same, instead of throwing their garbage into their neighbours'

backyard. ... The choice to turn Liossia into a dumpsite was a *CLASSIST ONE.* Let not this environmental 'apartheid' go on indefinitely. (Pappous, Mayor of Liossia, 2008, emphasis original)

Conclusions

We have analysed the grassroots mobilisations against landfill siting in three municipalities of Attica by adopting the following sequential steps. We started by acknowledging, through a review of the relevant literature, that the opposition to LULUs is a phenomenon shared across different national contexts. We continued by identifying the two frames which, for a variety of intents and purposes, are used to label these types of mobilisations, the NIMBY and environmental justice frames. We proposed that both frames can come together around two key dynamics that they share: the risk and lack of trust in expert authorities that appear to engulf the claims put forward by affected communities. As such, local opposition against LULUs can be located within the tensions of late modernity, as articulated in the works of Ulrich Beck and Anthony Giddens. However, in the light of the well entrenched discussions of the 'Mediterranean Syndrome' (La Spina and Sciortino 1993) and the 'Southern Problem' (Pridham and Cini 1994), we recognised that these mobilisations could very well be seen as typically Southern and/or Greek instead of ad hoc mobilisations characteristic of late modernity. Thus, we proceeded to a presentation on the state of municipal waste management in Greece in general and the arguments surrounding the siting of HETAs in the greater Athens region in particular. There we identified the interrelationship between the noticeable inefficiency exhibited by the institutional apparatus (resulting from poor planning, inadequate funding and fragmentation) and the emergence and consolidation of intense local opposition to the development of new, well planned and managed landfills that HETAs purport to be. This set the framework for the analysis of the protest events surrounding the siting of HETAs. For this analysis we used data found in the online version of the Greek daily *Eleftherotypia* and on the web pages of groups involved in the protest.

Mobilisations against waste treatment facilities ran the risk of being all too easily written off as 'NIMBY' reactions, implying an 'irrational' and or 'short-sighted/selfish' behaviour on behalf of local communities, which wish to reap the benefits of a certain development yet try to avoid bearing their share of the burden. Previous research has demonstrated that easy generalisations can be misleading. Local inhabitants mobilising against LULUs can do so for a variety of reasons, in which selfish NIMBY calculations are interwoven with issues of equity, fairness, and lack of trust in state officials and scientists.

Our analysis of the Greek case has confirmed that this is indeed the case. Although NIMBY concerns were evident, what seems to have made the locals tick is what they perceived to be the state's indifference to sound 'scientific' evidence and their exclusion from the deliberation process. These two factors contributed to their long-standing mistrust of the (state-regulated) planning process. This is particularly evident from the fact that they kept on equating (although by dubious means) the proposed HETAs with the notorious landfills. As one blog entry reads, the locals '*know* what a HETAs is and how it operates in reality in Greece'. And they actually did have such 'experience': for almost 20 years the citizens of Attica, and the whole of Greece for that matter, have been well accustomed to the appalling conditions of operation and the string of broken promises to local inhabitants concerning the only existing landfill, at Liossia, western Attica. With that 'common knowledge' in store, it is hardly surprising that the affected communities were particularly wary of these 'new' waste disposal sites.

Things were not made any easier by the fact that the mobilisers were convinced that they had 'Science' on their side. We have shown that the affected communities question the scientific soundness of establishing HETAs in their area, but the developers had yet other studies, supporting the original planning. This is not surprising, since these different studies have used different threshold criteria, and thus have reached different conclusions. However, in our view, the most important development has been the 'politicisation' of the scientific debate. The reader should recall that northern Attica's major scientific argument has been to 'take the waste out of Attica'. This approach, questionable in terms of equity, has indeed been the one advocated by most of the scientific teams involved. Yet successive Ministers for the Environment have rejected that scientific suggestion, although they have never quite explained why they did so. The fact remains that what came to be labelled as 'the dogma of non-export', by scientists and journalists alike, gave to the opposing communities a reason to question the planners' commitment to sound scientific principles, and to suspect that the proposals were largely predefined.

We also found that the discourse of opposition to the siting differs across the affected communities. In eastern and northern Attica, to date unaffected by waste management installations, we find the whole range of arguments. By contrast, in eastern Attica, it is issues of fairness and equity that set the tone: locals opposing the proposed development make a strong case in terms of the 'unfairness' of their being expected to host HETAs when they have suffered for so many years because of the existing landfills.

Since the public's mistrust of scientific claims made by state authorities and developers, and concerns about the lack of democratic accountability, were omnipresent, it is very tempting to consider these mobilisations as manifestations of the tensions characteristic of late modernity. However, at the same time, we should not lose sight of the fact that the particulars of these mobilisations are always the result of a filtering process with much older historical roots. As such, although we can encounter the phenomena under examination in many national contexts, the negativities associated with the Greek policy process give them a special flavour. Further cross-national study of ad hoc mobilisations will be needed before we can safely align with the grand social theorisation characterising the reflexive modernisation perspective.

Notes

1. See Hubbard (2005) for an interesting discussion of such processes concerning local opposition against an asylum centre.
2. The Kouroupitos judicial saga begun in 1988 and the first ECJ ruling was issued in 1992. Greece had paid a total fine of €4.72 million up to February 2001, when the Commission was satisfied with the site's closure.
3. Which is under the responsibility of municipalities and communities. The reader should note that as late as 1997 there existed approximately 6000 municipalities and communities, each responsible for its own waste management arrangements. The Greek local government's reshuffling of that year, bringing the number down to a little over 1000 OTAs (Organisations of Local Self-government), allows for a new approach where the existing landfills (largely servicing a single town or village) will be replaced with larger, and collectively managed HETASs accommodating several OTAs.
4. Note that not all of Attica's municipalities and communities participate in ESDKNA. Most notably, those of north-eastern Attica do *not* and thus are not (allowed to) deposit their waste at Liossia landfill.
5. Note that currently the area's waste is disposed over around a dozen of different landfills and dumping sites, some of them illegal.
6. See newspaper *Eleftherotypia*, 14 July 2008
7. In the mass media even the operating HETAS in eastern Attica is referred to as 'landfill', the two terms used largely indiscriminately.
8. The only relevant site, that of the 'Trans-municipal Committee for removing the Liossia landfill', was active only for the second half of 2005 (http://geocities.com/diadimotiki_epitropi/).

References

Ackerman, F. and Mirza, S., 2001. Waste in the inner city: asset or assault? *Local Environment*, 6 (2), 113–120.

Agapitidis, I. and Frantzis, I., 1998. A possible strategy for municipal solid waste management in Greece. *Waste Management and Research*, 16 (3), 244–252.

Almond, G. and Verba, S., 1963. *The civic culture*. Princeton, NJ: Princeton University Press.

Andreou, G., 2004. *Multilevel governance: EU landfill directive in Greece*. OEUE Phase II, Occasional Paper 4.5–09.04.

Anti-HETAs, 2008. What is the HETAS? *Anti-HETAS blog*. Available from http://antixyta.blogspot.com/2008/07/blog-post_18.html [Accessed 28 July 2008].

Apostolopoulos, M., 2008. A deadly trap set -and I think you have all stepped in it. *Oust the HETAS blog*. Available from http://exotaxyta.blogspot.com/2008_07_01_archive.html#3046280087863833236 [Accessed 28 July 2008].

Arkoudaris, C., 2008. Stop this injustice done to us. Address to a conference organised by Attica's Federation of Municipality, Prefecture of Western Attica press release, 7 February. Available from http://www.nada.gr/index.php?option=com_content&task=view&id=382&Itemid=230, [Accessed 28 July 2008].

Bacot, H., Bowen, T., and Fitzgerald, M.R., 1994. Managing the solid waste crisis. exploring the link between citizen attitudes, policy incentives, and siting landfills. *Policy Studies Journal*, 22 (2), 229–244.

Baxter, J., Eyles, J., and Elliott, S., 1999. 'Something happened': the relevance of the risk society for describing the siting process for a municipal landfill. *Geografiska Annaler*, 81B (2), 91–109.

Beck, U., 1992. *Risk society: towards a new modernity*. Translated by Martin Ritter. London: Sage Publication.

Beck, U., 1995. *Ecological politics in an age of risk*. Translated by Amos Weisz. Cambridge: Polity Press.

Botetzagias, I., 2006. Local mobilizations against waste disposal sites in Greece. *In*: *Paper presented at the ECPR Joint sessions, workshop 'Local Environmental Mobilizations'*, 25–30 April, Nicosia.

Buclet, N. and Godard, O., 2001. The evolution of municipal waste management in Europe: how different are national regimes? *Journal of Environmental Policy and Planning*, 3 (4), 303–317.

Burningham, K., 2000. Using the language of NIMBY: a topic for research, not an activity for researchers. *Local Environment*, 5 (1), 55–67.

Capek, S.M., 1993. The 'environmental justice' frame: a conceptual discussion and an application. *Social Problems*, 40 (1), 5–21.

Cavatassi, R. and Atkinson, G., 2003. 'Social' and 'private' determinants of opposition to landfill siting in Italy. *Journal of Environmental Assessment Policy and Management*, 5 (1), 27–43.

Dalamagka, T., 2008. Local reactions against HETAS sitings: the cases of Athens and Thessaloniki. Unpublished thesis (MSc), Department of Environment, University of the Aegean [in Greek].

Davies, A., 2007. A wasted opportunity? Civil society and waste management in Ireland. *Environmental Politics*, 16 (1), 52–72.

European Environment Agency (EEA), 2007. Europe's environment – The fourth assessment. Available at http://www.eea.europa.eu/publications/state_of_environment_report_2007_1 [Accessed 13 October 2009].

European Environmental Agency (EEA), 2008. *Municipal waste generation in western European (WE) and central and eastern European (CEE) countries*. Available at http://themes.eea.europa.eu/IMS/IMS/ISpecs/ISpecification20041007131809/IAssessment1183020255530/view_content [Accessed 13 October 2009].

Economopoulos, A., 2007. *Managing domestic waste: problems of the national planning and solutions*. Available from http://www.oxistoxyta.gr/docs/Hmerida Oikonomopoulos.doc [Accessed 28 July 2008].

Elafros, Y., 2005. Garbage: the landfills of political promises. *Kathimerini* newspaper, 9 July. Available from http://www.kathimerini.gr/4Dcgi/4dcgi/_w_articles_kathcommon_1_09/07/2005_1284281 [Accessed 28 July 2008].

Eleftherotypia, 1996. Garbage: 1000 tonnes per day and the landfill is closed. *Eleftherotypia*, 6 June. Available from http://archive.enet.gr/1996/06/06/on-line/keimena/greece/greece2.htm [Accessed 28 July 2008].

Freudenburg, W.R. and Pastor, S.K., 1992. NIMBYs and LULUs: stalking the syndromes. *Journal of Social Issues*, 48 (4), 39–61.

Giddens, A., 1990. *The consequences of modernity*. Cambridge: Polity Press.

Giddens, A., 1991. *Modernity and self-identity: self and society in the late modern age*. Cambridge: Polity Press.

Harris Ali, S., 1997. Trust, risk and the public: the case of the Guelph landfill site. *Canadian Journal of Sociology*, 22 (4), 481–504.

Hatziioannidou, E., 2008. Huge fine for the landfills. *Kathimerini* newspaper, 8 July. Available from http://news.kathimerini.gr/4dcgi/_w_articles_ell_100030_02/07/2008_276152 [Accessed 28 July 2008].

HERRCO, 2007. Newsletter No. 9, April. Available from http://www.herrco.gr/ConDows/Default/8_newsletter_April07.pdf [Accessed 28 July 2008].

Hubbard, P., 2005. Accommodating otherness: anti-asylum centre protest and the maintenance of white privilege. *Transactions of the Institute of British Geographers*, 30 (1), 52–65.

Hubbard, P., 2006. NIMBY by another name? A reply to Wolsink. *Transactions of the Institute of British Geographers*, 31 (1), 92–94.

Hunter, S. and Leyden, K.M., 1995. Beyond NIMBY. Explaining opposition to hazardous waste facilities. *Policy Studies Journal*, 23 (4), 601–619.

Kasperson, R.E., 2005. Siting hazardous facilities: searching for effective institutions and processes. *In*: S.H. Lesbirel and D. Shaw, eds. *Managing conflict in facility siting. An international comparison*. Cheltenham: Edward Elgar, 13–35.

Kelepertzis, A., 2003. *Environmental impacts of the proposed HETAS at the area 'Mavro Vouno', Grammatiko, northeastern Attica*. Department of Geology, University of Athens. Available from http://www.oxistoxyta.gr/docs/university1.pdf [Accessed 28 July 2008].

Kitschelt, H., 1986. Political opportunity structures and political protest: anti-nuclear movements in four democracies. *British Journal of Political Science*, 16, 57–85.

Kollias, G., 2008. *HETAS or HETES? Rubbish*. Youtube video uploaded on 14 April. Available from http://www.youtube.com/watch?v=G1YVJteBkO8 [Accessed 28 July 2008].

Kolliou, S., 2008. TV interview for the show 'Reportage without frontiers: Greece –a huge landfill', 19 June. Available from http://xyta-lefkimis.blogspot.com/2008/06/blog-post_22.html [Accessed 28 July 2008].

Koukis, N., 2007a. Press release, 6 November. Available from http://www.oxistoxyta.gr/docs/deltio1.doc [Accessed 28 July 2008].

Koukis, N., 2007b. Press release – interview, 6 November. Available from http://www.oxistoxyta.gr/docs/deltio2.doc [Accessed 28 July 2008].

Kousis, M., 2007. Local environmental protest in Greece, 1974–94: exploring the political dimension. *Environmental Politics*, 16 (5), 785–804.

La Spina, A. and Sciortino, G., 1993. Common agenda, southern rules: European integration and environmental change in the Mediterranean states. *In*: J.D. Liefferink, P.D. Lowe, and A.P.J. Mol, eds. *European Integration and Environmental Policy*. London and New York: Belhaven Press, 211–236.

Lober, D.J., 1993. Beyond self-interest: a model of public attitudes towards waste facility siting. *Journal of Environmental Planning and Management*, 36 (3), 345–363.

Lober, D.J., 1995. Why protest? Public behavioral and attitudinal response to siting a waste disposal facility. *Policy Studies Journal*, 23 (3), 499–518.

Lober, D.J. and Green, D.P., 1994. NIMBY or NIABY: a logit model of opposition to solid-waste-disposal facility siting. *Journal of Environmental Management*, 40 (1), 33–50.

Maloutas, T., 2000. *The cities: social and economic atlas of Greece*. EKKE: Athens (in Greek).

McClymont, K., 2008. 'We're not NIMBYs!': Contrasting local protest groups with idealised conceptions of sustainable communities. *Local Environment*, 13 (4), 321–335.

Pappous, C., 2008. Oust the garbage now: we are not Attica's castaway children. Address to a conference organised by Attica's Federation of Municipality, Press release of February 7. Available from http://www.dimosliosion.gr/Articles.aspx?LangID=1&FolderID=0b0afba1-115f-43a5-ad72-2fc30fcd1f56&year=&searchstr=&PageNo=8&EntityID=603cd2a5-3b17-45a7-8c81-5ff05b20847d [Accessed 28 July 2008].

Pellow, D.N., 2000. Environmental inequality formation. toward a theory of environmental injustice. *American Behavioral Scientist*, 43 (4), 581–601.

Pridham, G. and Cini, M., 1994. Environmental standards in the European Union: is there a southern problem? *In*: M. Faure, J. Vervaele, and A. Weale, eds. *Environmental standards in the EU in an interdisciplinary framework*. Antwerp: Maklu, 251–277.

Rootes, C., ed., 2003. *Environmental protest in Western Europe*. Oxford University Press.

Rootes, C., 2007. Acting locally: the character, contexts and significance of local environmental mobilisations. *Environmental Politics*, 16 (5), 722–741.

Schively, C., 2007. Understanding the NIMBY and the LULU phenomena: reassessing our knowledge base and informing future research. *Journal of Planning Literature*, 21 (3), 255–266.

Selamazidis, M., 2008. Either the HETASs at Keratea and Grammatiko open or the Fili [HETAS] closes. *Eletherotypia* newspaper, 9 July. Available from http://www.enet.gr/online/online_text/c=112,dt=09.07.2008,id=91323072 [Accessed 28 July 2008].

Watson, M. and Bulkeley, H., 2005. Just waste? Municipal management and politics of environmental justice. *Local Environment*, 10 (4), 411–426.

Wolsink, M., 2006. Invalid theory impedes our understanding: a critique on the persistence of the language of NIMBY. *Transactions of the Institute of British Geographers*, 31 (1), 85–91.

Wright, S., 1993. Citizen information levels and grassroots opposition to new hazardous waste sites: are NIMBYists informed? *Waste Management*, 13 (33), 253–259.

Wynne, B., 1996. May the sheep safely graze? A reflexive view of the expert-lay knowledge divide. *In*: S. Lash, B. Szersynski, and B. Wynne, eds. *Risk, environment and modernity: towards a new ecology*. London: Sage Publications, 44–83.

YPEHODE (Ministry for the Environment), 2005. *Synopsis of the final report concerning the restoration of landfills*. Ministry for the Environment, Athens. Available from http://www2.minenv.gr/press/doc/0505252.doc [Accessed 28 July 2008].

Movements, mobilities and the politics of hazardous waste

Su-Ming Khoo and Henrike Rau

School of Political Science and Sociology, National University of Ireland, Galway, Ireland

Global flows of hazardous waste and waste management technologies are major sources of environmental contestation. They reflect political structures and struggles within, and between, developed and less developed countries. The 'new mobilities paradigm' is tested using two cases of protest in Malaysia and East Germany. Focusing on the conjunctures of various (im)mobilities, the ways in which political circumstances combine with the materialities of wastes and technologies are shown to affect the trajectories and outcomes of environmental protest. This challenges assumptions that mobilities of objects, people and ideas inevitably undermine governmentality. While the merits of 'mobilities' as a lens for inquiry are acknowledged, greater attention should be paid to the politics of (im)mobilisations.

Introduction

We consider the mobilisation of technology, capital and waste in the context of waste infrastructure development in Malaysia and East Germany and examine people's responses to the arrival of technologies and wastes in their localities. Contrasting the relative mobilities of things with the relative immobilities of people and place, we will specify particular conjunctures of *(im)mobilities* that constitute key sites of environmental contestation. We draw upon the 'new mobilities paradigm' (e.g. Urry 2000, 2007, Cresswell 2006) and show its relevance to the study of environmental protest, but also problematise, extend and politicise its current scope. Our analysis of environmental contestation centres on the complex interactions between the specific materialities of waste and socio-cultural and political changes.

The mobilities paradigm raises questions about 'too little movement for some or too much for others or of the wrong sort or at the wrong time' (Urry 2007, p. 6). According to Urry (2007, p. 11), 'analysing ... mobilities involves examining many consequences for different people and places that can [be] said to be in the fast and slow lanes of social life. There is a proliferation of places, technologies and "gates" that enhance the mobilities of some and reinforce the immobilities of others'. Laws and state policies regulating hazardous waste might be considered an example of such 'gates' which both enable and constrain the movement of waste. At the same time, these mobilities act as catalysts for collective responses when mobile waste and waste technologies arrive in particular localities.

While the focus on mobilities offers significant opportunities to re-think the politics of waste, more emphasis should be placed on (im)mobility, (in)equality and (im)mobilisation and their inherently political and contested nature. The substantive approach pursued here treats mobilities as a novel analytical perspective which complements existing work on social movements rather than a new paradigm. It asks what *kinds of (im)mobilities* are involved in environmental contestation and protest. Our examination of 'negative mobilities' centres on the particular nature of hazardous wastes and the impossibility of their disposal. Their impact on people and places raises political and ethical questions about the distribution of environmental burdens and consequences when mobile wastes are immobilised. A politicisation of the mobilities paradigm follows necessarily from such an analysis of the material dimensions of mobilities and their social consequences.

The two historic cases – Morsleben in East Germany and Bukit Nenas in Malaysia – draw attention to the limits of the 'new mobilities paradigm' by highlighting East–West and South–North flows. This opens up a 'view from the slow lane' of local environmental protest and its national and global connections and moves the analysis beyond a Western or Eurocentric focus. The case studies illustrate how structural transformations such as German unification and 'Asian Tiger' economic growth changed the economic and political landscape of these countries and profoundly affected the structures of environmental contestation.

In what follows, we will focus on the complexities surrounding community mobilisation in opposition to hazardous waste management facilities. An initial discussion of the merits and limitations of the 'new mobilities paradigm' for the analysis of environmental contestation in section two will be followed by an in-depth exploration of our two cases. In section five we compare and contrast mobilities and (im)mobilisations in the two cases. In conclusion we will argue for a greater emphasis on the materialities and politics affecting interdependent mobilities in social scientific investigations of environmental protest and green political movements.

Merits and limitations of the 'new mobilities paradigm'

Material and virtual mobilities – global flows of people, goods and information – are transforming the subject matter of (Western) sociology, that is, societies as territorially defined nation-states which effectively regulate these flows. For example, strategies deployed by individual nation-states for (not) dealing with waste materials result in the global circulation of materials, money and disposal technologies (e.g. shipping hazardous waste or transferring environmental technologies to poorer regions). However, these flows end somewhere, presenting examples of *problematic immobility* that both reflect and reproduce global patterns of social and environmental injustice. What is driving the movement of technology and where does the waste stop? How does it become embodied in those communities that are the recipients of waste and waste management technologies? What kinds of community mobilisation arise in response, and what kinds of material and political limits are placed on these responses? Finally, to what extent can the 'new mobilities paradigm' capture these (im)mobilities and the ways that they are shaped by power, coercion and consent?

Urry (2007, p. 9) uses metaphors of mobility as the vehicle for developing a 'sociology beyond societies'. His conceptual response to an increasingly globalised world is to focus on movement, mobility and contingency and attempts to move social theorising beyond stasis, structure and social order. Urry (2000, p. 1 ff.) advances a 'new paradigm' for sociology based on the idea of 'diverse mobilities of peoples, objects, images, information and *wastes*; and of complex interdependencies between, and social consequences of, these diverse mobilities' (emphasis added).

In contrast, Bauman (2004, p. 69) also uses the notion of 'waste' as a powerful metaphor, but for 'human waste ... or wasted humans', describing the irreducible segment of human society that can be neither accommodated nor moved: the human casualties of globalisation. In the past, power differentials between modernised and underdeveloped regions functioned as a 'safety valve', protecting the modernising regions by using the rest of the planet as a dumping ground for their toxic waste. Globalisation has blurred the division between centre and periphery as 'the global spread of the modern way of life has now reached the furthest limits of the planet' (Bauman 2004, p. 69). Bauman's (2004, p. 5) assertion that '[t]he planet is full' is thus not a *technical* statement, but a *sociological and political* one about the human costs of globalisation and the inevitability of contestation. Here we illustrate Bauman's analysis of the human costs, by focusing on hazardous wastes which pose particular dangers to human bodies.

Urry (2000, 2007) highlights the de-politicising consequences of globalisation's new mobilities. He suggests that social governmentality, that is, *politics*, is undermined by mobilities of objects, bodies, representations, virtual technologies and information flows (Urry 2007, p. 47). His 'new mobilities paradigm' emphasises the centrality of movement, travel and communication. His interest in the imaginative and virtual realms downplays concrete and embodied materialities as a source of contention. We use examples of hazardous waste management and protest from Malaysia and East Germany to show that politics is both *source* and *product* of concrete (im)mobilities. Politics, like waste, can be mobilised or displaced, but does not go away. This confirms Rootes' (2007, p. 722) contention that local campaigns 'serve as reminders ... that environmental issues have not been quietly absorbed by bureaucratic administration and representative democratic politics but remain as matters of fundamental contention'. This clearly challenges the view that political structures matter much less in an increasingly mobile world. The western liberal assumptions of the mobilities paradigm are therefore undermined as we reframe the mobilities of environmental goods and bads as problems of contestation. Urry's (2007, p. 47) five interdependent mobilities thus not only produce and shape social life, they also are intrinsically political.

Reconstructing the social as mobility provides a useful entry point to the study of waste and its contestation. Our cases show that rapid changes to material realities as well as political structures can serve to either fuel or demobilise political action and protest, sometimes in unpredictable ways. The mobilities perspective highlights the processual and contingent character of politics which contrasts with the depiction, in social movements literature, of collective mobilisation as a derivative of fixed political structures. We concur with Rootes (1997) that many aspects of political context can be relatively unstable and contingent (see also Rucht 1990). Mobilities of people and technologies may drive the emergence of new structures of collective action, as is exemplified by NGO activity in Malaysia and East Germany. Our two examples of (semi) authoritarian states show that political structures may *immobilise* civil society through surveillance, restrictions on movement and civil liberties (compare Foweraker 1997). However, even these immobilities are subject to change, as their effects can rebound on society and the political system, leading to significant transformation.

We illustrate and expand the mobilities analysis by focusing on transboundary and global trade in hazardous waste. While the 'new mobilities paradigm' tends to privilege the transcendence of territory and spatial boundaries, the politics of mobilisation highlights the importance of material reality, locality and nationality, even in the context of globalisation. Disputes over the (un)just (im)mobilisation of hazardous waste exemplify the persistence of local protests in the face of transnational waste flows and the growing regulation, bureaucratisation and eco-modernisation of waste disposal. The cases presented in the following sections of this paper – the Bukit Nenas hazardous waste facility in Malaysia and the Morsleben nuclear waste repository in East Germany – illustrate these limits to 'problem displacement' They also emphasise the importance of the particularities of place, culture and identity in the mobilisation of local resistance to hazardous waste disposal facilities.

Hazardous waste, community mobilisation and economic growth in Malaysia

The generation and treatment of hazardous waste are contentious issues that form key focal points for the global environmental justice movement, especially when wastes and technologies are seen to flow in one direction – from developed countries in the 'North' to less developed ones in the 'South'. Hazardous waste incineration has played a key role in both transnational environmental mobilisation and local protests around the world. Concerns about Persistent Organic Pollutants (POPs), a by-product of waste incineration, link anti-incinerator protests in Malaysia to long-standing international campaigns about toxics and pesticides. POPs represent mobility and immobilisation as they are transmitted through milk and accumulated in tissue. The Stockholm Convention on POPs lists 12 main types of POPs (the 'dirty dozen'). These include dioxins, furans, Polychlorinated Biphenyls (PCB), Hexachlorobenzene (HCB) and the pesticides dieldrin and DDT which became notorious following the publication of Rachel Carson's bestseller *Silent Spring* in 1962. How have political and institutional structures, historical circumstances and mobilities (virtual and material) combined to influence environmental mobilisation in Malaysia against these pollutants? The discussion in this section uses published documents and fieldwork conducted in the Bukit Nenas area, with the assistance of Consumers Association of Penang (CAP) activists.

Background: economic development and the demobilisation of Malaysian civil society

The Malaysian context is one of rapid economic development, population growth, and structural transformation. At independence in 1957, Malaysia was a primary commodity producer with a relatively small, mainly rural population of 7 million. Today Malaysia is industrialised, with manufactures accounting for over 75% of exports. Its predominantly urban population now exceeds 27 million. It has practised a 'develop first, clean up later' model of development. However, as environmental pressures have increased, authoritarian 'brown modernisation' has given way to a pragmatic and market-led version of 'green modernisation'.

Economic liberalisation measures are often credited as the engine of growth and structural transformation in Malaysia (Salleh and Meyanathan 1993). From the late 1980s, Malaysia adopted liberalised export-oriented industrialisation policies, with a drive to create 'Malaysia Inc.' through privatisation (Standing 1993, p.42). The liberalisation of finance and investment was accompanied by efforts to deepen domestic participation and localise industry, effectively blurring the distinctions between foreign and domestic and between state and corporate interests. Liberalisation means that much of the responsibility for environmental protection is devolved to private corporations, such as Kualiti Alam (see below). The latter is lauded by the government as an example of 'green capitalism', fostering eco-efficiency and corporate social responsibility.

During the 1980s and early 1990s Malaysian civil society was weak and ethnically divided (Loh and Khoo 2002, Verma 2004) and presided over by a semi-authoritarian 'repressive-responsive regime' which became increasingly hostile to NGOs (Crouch 1996). The Malaysian state co-opted, repressed or demobilised any political activities which it considered to be potentially oppositional. Authoritarian measures were introduced in 1970 when constitutional amendments restricted civil and political liberties, including freedom of speech, freedom of information and freedom of assembly. This authoritarian tendency increased under Mahathir's government in the 1980s, culminating in mass arrests of community activists and politicians in 1987 (see Means 1991, International Bar Association *et al.* 2000, Verma 2004). In response, NGOs built coalitions to oppose the increasing authoritarianism of the executive and the politicised and nepotistic awarding of contracts to favoured companies (Means 1991, Gomez 1994, Gomez and Jomo 1999). Environmental campaigning mainly centred around large infrastructural projects and the development of environmental technologies such as waste incineration.

Despite these constraints, Malaysian environmental NGOs developed considerable capacity to mobilise campaigns. Some, like CAP and SAM (*Sahabat Alam Malaysia*, Friends of the Earth Malaysia) became professionalised and internationalised since the early 1980s, which saw the rise of a core group of expert and outward-looking Malaysian activists with a global perspective. Malaysian environmental NGOs illustrated the 'boomerang effect' (Keck and Sikkink 1998), whereby domestic repression or passivity forces the internationalisation of local campaigning. By linking with international networks, Malaysian NGOs were able to mobilise information, expertise and funding to increase their standing and credibility, rebounding on the national context as international leverage.

A brief history of CAP, anti-toxics campaigns and the Bukit Nenas waste facility

One of the most active and established environmental organisations in Malaysia is CAP and its sister organisation, SAM. Since the late 1970s, CAP/SAM has been an important national and regional actor in global environment, development and anti-toxics campaigns. CAP combines strong grassroots campaigning experience with a broad-based approach that integrates consumer, development and environmental issues. These include public and occupational health, indigenous people's survival, citizen rights, education and protest against large-scale projects such as dams and incinerators. It also provides public information about incineration, toxicity and pollution, within a wider context of product safety, testing and awareness-raising.

CAP/SAM works with community groups such as farmers, plantation workers and villagers, as well as in a more structured fashion with schools, government departments and employers. Its relationship with government ranges from critical collaboration to radical opposition. CAP's independent stance, bottom-up approach and explicitly *Southern*, 'value for people' orientation contrast with the 'value for money' approach pursued by many Northern consumer associations (Fazal 1982). It uses consumer protection issues to mobilise people, with the aim of changing the model of development towards more people-centred, participatory and sustainable approaches, and provides expert and legal advice to community groups to contest cases.

From the late 1970s, CAP/SAM became involved in international toxics campaigns through work with farmers and plantation workers on dangerous pesticides. In the 1980s and early 1990s, Malaysian anti-toxics campaigning predominantly centred around an emerging Southern consensus about the problem of toxic 'dumping' – the export of waste and outdated, polluting technology from the North to the South (e.g. Third World Network 1988, UNEP 2004). The 1992 Basel Convention enshrined this consensus by attempting to immobilise transboundary waste exports. However, poor countries are left with the problem of safe disposal for existing hazardous waste. This can lead to a scenario where the waste is immobilised in the poor country with no indigenous capacity for treatment developed, while local contention is immobilised through neglect or repression.

In 1979, the Malaysian government began to take a more 'green modernisation' approach and initiated legislation governing hazardous or 'Scheduled Waste'. In the 1980s, a potential site for a hazardous waste treatment facility had been identified within rubber plantation lands at Bukit Nenas in rural Negeri Sembilan. In the late 1980s liberalisation and privatisation policies meant that the facility would most likely be developed by a private concessionaire. Government legislation, in the form of the Scheduled Waste Act 1989 was drafted in consultation with business interests. The legislation was modelled on Danish legislation, with technical assistance from Danish Cooperation for Environment and Development (DANCED). The Bukit Nenas plant was modelled on the Kommunekemi project in Nyborg, Denmark, cited by international environmental activists at the time as an example of 'good practice' in separation and treatment of hazardous wastes (Meadows 1991, p. 185). In 1991 Kualiti Alam, a Danish–Malaysian consortium, was awarded a 15-year concession to build and operate the hazardous waste separation, treatment and incineration facility, at a cost of US$70 million.

CAP's objections to the incinerator at Bukit Nenas centred around flaws in the Environmental Impact Assessment, human health and environmental safety concerns, especially dioxin emissions and the potential for groundwater contamination. They also criticised the lack of public accountability and consultation surrounding the project and the lack of discussion of any alternatives to incineration. Its stance was influenced by its wider links to international environmental and health activism on toxics, incineration, pesticides, and baby milk. CAP estimated that the proposed plant would directly affect 1879 families from seven Chinese New Villages and five Malay villages, with a further warning that up to 50,000 people in the district could be affected. After considerable awareness-raising and guidance from CAP, Malay and Chinese villagers formed an Anti-Toxic Waste Committee in 1991. Pressure from the campaign led to a 'public meeting' with the Minister for Science, Technology and Environment, but there was no opportunity for public participation and the frustrated residents resorted to picketing, but the protest was muted due to the presence of riot police.

Mobilisation against the incinerator at Bukit Nenas stagnated as the project did not proceed until 1995. By this time, the anti-toxic waste committee formed in 1991 had been substantially demobilised in various ways. Active committee members had lost their positions on their respective village committees. Discussions between Malay and Chinese village heads had become politicised in other ways, and the members felt reluctant to act decisively as they feared that mobilising opposition to the incinerator might unintentionally inflame underlying ethnic tensions. The affected villagers included Chinese pig farmers, whose farms were sources of ethno-religious contention.[1] The Chinese community leader who served as Chairman of the Anti-Toxic Waste Committee was also understandably reticent since he had previously been imprisoned without trial under Malaysia's draconian Internal Security Act, and feared reprisal if he was branded as an 'anti-development agitator'.

Concepts of economic progress and 'development' were used to demobilise the protest at Bukit Nenas by reframing the project as 'bringing development'. The Malay villagers were encouraged to individually 'participate in development' by acquiring shares in Kualiti Alam as part of the government's affirmative action scheme to promote Malay ownership of corporate equity (see Gomez and Jomo 1999). The rezoning of agricultural land into industrial land led to the co-optation of the main landowner, the rubber plantation company, Guthrie, as well as some landowning villagers. Guthrie initially opposed the project from 1991 to 1993 but by 1995 the company agreed to support the project. Construction of the project proceeded in earnest after a controlling stake in Kualiti Alam was bought by the powerful, ruling party-dominated conglomerate, United Engineers Malaysia (UEM), meaning that powerful political–business interests were now behind the project (Gomez 1994, p. 90–94).

Since the late 1990s, capital and technology flows have displaced authoritarianism as the major forces immobilising protest around the Bukit Nenas site. This is partly due to the necessity created by increasing quantities of hazardous waste, which are mobilised and accumulated by high-growth industrial development. Capital-intensive, high-technology solutions are increasingly preferred by Malaysia's developmental state and the politically powerful concessionaires who have won contracts to provide the means to reduce and immobilise waste problems. As more hazardous waste is generated nationally, waste management becomes both necessary and highly profitable. Kualiti Alam now claims that it provides one of the most comprehensive waste management facilities in Southeast Asia. As the Bukit Nenas plant has now been working for a decade, it is recognised that that it provides the 'right way' to dispose of hazardous waste, in comparison with illegal dumping, which is emerging as a worse option (Cruez 2006).

Between 1987 and 1994 when the Kualiti Alam facility at Bukit Nenas became fully operational, 125,000 tonnes of toxic waste had already accumulated for disposal. According to the Basel Convention country reports for Malaysia, amounts of scheduled waste more than doubled between 2003 and 2006. About 15% of this scheduled waste was imported from abroad in 2006, and multinational operations in Malaysia accounted for 40% of the waste processed by Kualiti Alam. Although transboundary movement is somewhat significant, the major issue is not that Malaysia is acting as a pollution haven for 'Northern' toxic dumping, but that it is generating ever larger quantities of hazardous waste through its own indigenous economic development.

In upwardly-mobile industrialising countries like Malaysia, the differences between North and South have blurred. Since the later 1990s, the debate has moved towards a partial acceptance of 'clean technologies', compliance with international agreements and the development of appropriate Southern environmental expertise (e.g. Third World Network 2001). Liberalisation and privatisation have driven a corporate view of waste disposal as a profitable industry in itself, with increasing emphasis on corporate social responsibility within a privatised market model.

The 'technology transfer' advocated by Southern NGOs and governments has undoubtedly taken place to a significant extent. High-tech industrialisation and technology transfer objectives have dominated the 'Malaysia Inc.' vision, leading the country to invest in developing industrial capability in incineration. Malaysia is a relatively large importer of high-tech incineration plant. However, by 2007, Kualiti Alam was aiming to provide 70% of the design and development of a range of incinerators, with 30% technology transfer and design certification provided by Danish engineers (Bernama 2007).

NGOs like CAP continue to advocate against industry-driven high-tech waste management, arguing instead for an alternative approach that focuses on waste reduction and the elimination of toxics. They emphasise the precautionary principle, pointing to the fact that pollutants are produced even by improved incineration technology, and these pollutants are difficult to monitor (Greenaction/Gaia 2008). NGOS can continue to play an important role in developing expertise, monitoring and compliance. Since 2004, CAP has been involved in the International POPs Elimination Project (IPEN) which monitors Persistent Organic Pollutants under the auspices of UNIDO (United Nations

Table 1.   Malaysia: Scheduled wastes produced, imported and exported.

	Scheduled waste produced (metric tons)	Scheduled waste exported	Scheduled waste imported
1987–1994 (accumulated)	125,000	No data	No data
1995–1999 (est)	431,000	No data	No data
2003	460,866	2,363	305,398
2006	1,103,456	5,806	172,151

Sources: Consumers Association of Penang 2005, Basel Convention Malaysia Country reports 2003 and 2006.

Industrial Development Organization) and UNEP (United Nations Environmental Programme).

The (im)mobilisation of radioactive waste: environmental protest in East Germany

The more recent history of German environmentalism has been inextricably linked to struggles over the (im)mobilisation of nuclear waste. Large-scale campaigns to stop CASTOR[2] transports to radioactive waste storage facilities in Gorleben and other key sites have come to symbolise grassroots resistance and environmental contestation in post-unification Germany (Blowers and Lowry 1997, Fischer and Boehnke 2004). Here we focus on the evolution of local protest in and around Morsleben, a former salt-mining village in East Germany, and the location of a final repository (Morsleben – ERAM) for low- and intermediate-level, non-heat-generating radioactive waste. It will address the question how political and institutional structures, historical circumstances *and* multiple material and virtual mobilities combine to create context-specific conditions for grassroots mobilisation and collective action. The changing politics of (radioactive) waste before and after German unification (late 1960s to early 2000s) and the resulting patterns of (im)mobilisation of people and waste form the backdrop to this instance of environmental contestation. Using published reports and original documents, including a detailed history of the ERAM facility (Beyer 2004, 2005) and web-based material, the account will highlight some of the specificities of East and West German environmental mobilisation which can only be understood if placed in their historical and political context.

Old burdens and lasting problems: a brief history of the Morsleben repository (ERAM)

Morsleben repository (ERAM, Endlager für Radioaktive Abfälle Morsleben) was previously a rock salt mine in Saxony-Anhalt near the former border between East and West Germany. It was chosen in the late 1960s by the GDR government (German Democratic Republic, 1949–1990) as the disposal site for low- and medium-level radioactive waste. The decision to open ERAM was connected to the construction of nuclear power plants in Rheinsberg (near Berlin) and Greifswald-Lubmin and a research reactor in Rossendorf near Dresden. This was indicative of the GDR's commitment to the development of nuclear technology for energy generation. During its service life from 1971 to 1998, ERAM received almost 37,000 cubic metres of both solid and liquid low- and medium-level radioactive waste from various sources (see Table 2). The first deliveries of radioactive waste arrived in Morsleben as early as 1971, 10 years prior to the completion and full licensing of the facility. Between 1971 and 1991 approximately 14,400 m^3 of radioactive waste and more than 6000 radioactive sources were stored in Morsleben (see Beyer 2004 for a detailed account).

Table 2.   Disposal of waste in Morsleben – sources and volumes.

Waste producer group	Volume in m^3
Nuclear power plants	23,816
Decommissioned nuclear power plants	6,528
Research institutions	2,592
Nuclear industry	159
State collecting facilities	3,090
Others	523
Reprocessing	45
Total	36,753

Source: Federal Ministry for the Environment, Nature Conservation and Nuclear Safety (BMU) 2006.

After German unification, ERAM was temporarily shut down in 1991. However, the then Minister for the Environment Klaus Töpfer invoked a special clause in the German Unification Treaty (*Einigungsvertrag*) to re-open the facility in 1992. This clause guaranteed the continued use of ERAM (and other East German industrial facilities) until 2000 (Beyer 2005). According to Greenpeace (2003), between January 1994 and September 1998, 22,320 m^3 of radioactive waste, mostly from West German nuclear and interim storage facilities, were brought to Morsleben. This included the surface-level interim storage of wastes which could be classified as high-level radioactive (e.g. containers with Caesium-137). The exact amount and location of radioactive waste within the ERAM facility remains subject to debate and there have been repeated finds of previously undocumented material (Beyer 2005, p. 20). Speculations that the GDR government may have accepted nuclear waste from West Germany prior to 1989 to generate income have not been confirmed by any reliable documentation.

The re-opening of ERAM led to local protests and disputes between environmental activists and the BfS (*Bundesamt für Strahlenschutz*, Federal Office for Radiation Protection in Germany, responsible for Morsleben and other nuclear waste repositories). Sustained pressure from local, regional and national environmental groups and legal action taken by Germany's largest environmental NGO, BUND (Bund für Umwelt und Naturschutz Deutschland, German branch of Friends of the Earth International) and Greenpeace led to the eventual closure of ERAM in 1998. More recently, the Morsleben facility has been the subject of renewed criticism and environmental protest after it emerged that the structural integrity of the central area of the mine was compromised, when debris fell on top of the waste containers (Greenpeace 2003). This also confirmed previous safety concerns, including those raised by East German scientists in the 1970s and 1980s (Greenpeace 2003, Lindemann 2007). Furthermore, a report published by Greenpeace in 1997 claimed that Morsleben posed a much greater threat to local and regional watercourses than previously anticipated (Greenpeace 1997). Plans are currently under way to permanently close and seal the ERAM site. Some environmental activists have challenged this decision, calling for greater transparency and an in-depth exploration of all possible closure options, including the retrieval of all wastes from the salt mine to avoid long-term radioactive pollution of watercourses. The planning process for the closure of ERAM is ongoing, and environmental campaigns continue to take place in and around Morsleben (see Lindemann 2007).

Crossing the border: German unification and the transformation of the environmental protest in Morsleben

What can the Morsleben case reveal about the political–institutional determinants of environmental contestation and their transformation over time? A direct comparison of local opposition to ERAM in East and West Germany before and after German unification in 1990 illustrates the importance of political systems and structures for the formation of environmental movements (see Beyer 2004 for a detailed and well researched account). Resistance to ERAM in the East German part of the Morsleben area was minimal due to lack of information. Historical documents reveal that most people in Morsleben knew very little about the scale of the ERAM repository and its inherent health and environmental risks. ERAM was subject to intense secret police activity (by the *Staatssicherheit* or *Stasi*), including reports on the overall mood of the public in the Morsleben area regarding the repository (Beyer 2004, p. 55, cf. Bastian 1996). ERAM's location near the inner-German border was the result of a strategic decision by the GDR authorities to facilitate and justify above-average levels of surveillance in this high-security area. The transportation of nuclear waste material to Morsleben was also strictly controlled. Everyday life in Morsleben prior to 1989 was thus shaped by various *immobilities* arising from state control and surveillance and this clearly limited opportunities for dissent and environmental protest.

In contrast, there was considerable disquiet among people in the Helmstedt area, a West German border town less than 5 kilometres away from Morsleben. During the 1980s, people in Helmstedt expressed their concerns about possible radioactive contamination of local wells caused by ERAM. An article published on 3 August 1987 in the regional West German newspaper *Hannoversche Allgemeine Zeitung* covered local concerns about potential water pollution and the lack of reliable information about ERAM. West German authorities unsuccessfully requested more detailed information on ERAM from their East German counterparts (Hänel 1998), and this information deficit further contributed to growing concerns in the Western part of the Morsleben area about potential health risks.

Rapid changes in political structures in 1989–1990 associated with the collapse of the Eastern bloc and German unification opened up considerable opportunities for environmental protest against the Morsleben facility. In November 1990 a citizens' initiative (*Initiative gegen das Atommüllendlager*

Morsleben) brought together concerned citizens from both sides of the former inner-German border to campaign for the immediate closure of ERAM. Established West German environmental groups based in Gorleben/Wendland (*Bürgerinitiative Umweltschutz Lüchow Dannenberg*), Salzgitter (*AG Schacht Konrad*) and other locations in the region that had previously been affected by radioactive and toxic waste issues extended their protest to Morsleben. In 1990 the *AG Schacht Konrad* and the *Bürgerinitiative Umweltschutz Lüchow Dannenberg* initiated events in Morsleben to inform and mobilise local people, including a 'Sunday walk' to the ERAM facility and a 'wedding-eve party' (*Polterabend*) on 2 October 1990 to draw attention to the 'unification' of Germany's radioactive waste problem. In November 1993, Greenpeace organised a large-scale event (*Morsleben Aktion Stillegung*) to highlight the dangers of the ERAM repository. In September 1995, a new environmental group – Greenkids e.V. – was set up in Magdeburg (capital city of Saxony Anhalt in East Germany, near Morsleben) and immediately initiated protests against the ERAM facility (Greenkids e.V. 2008). Greenkids activists also organised a mobile exhibition on Morsleben (*Morsleben – Geschichte eines umstrittenen Atomprojekts*), and collected, archived and published previously inaccessible information about ERAM (e.g. Beyer 2004). The formation of a regional network of anti-CASTOR groups and activists (*AntiCastorNetz Magdeburg*) marked another milestone in the development of environmental protest in the Morsleben region.

Sustained pressure and legal action taken by environmental groups eventually led to the closure of ERAM in 1998 and initiated heated debates about the future of the facility. The ERAM closure coincided with significant changes in the political landscape in Germany: the end of the 'Kohl era' in 1998. The new red–green coalition government of SPD and *Bündnis 90/Die Grünen* introduced legislation (*Atomkonsens*) to gradually phase out nuclear energy in Germany and to ban construction of any new nuclear power plants. However, serious problems regarding storage of radioactive waste remain. The search for a final storage facility (*Endlager*) in Germany continues and has led to environmental protests in Morsleben, Gorleben and other locations. Legal challenges and debates in the media about the merits and limitations of the *Atomkonsens* and the apparent revival of pro-nuclear arguments worldwide (Scally 2008) have accompanied these protests. Recent exports of German radioactive waste to La Hague (France) and Sellafield (UK) for re-processing have again highlighted the global dimensions of the radioactive waste problem.

The Morsleben case illustrates the significance of political and institutional structures for the emergence of a protest movement against hazardous waste facilities in Germany. It confirms Rucht's (2003) assertion that the very limited comparability of pre-1989 East and West German social movements can be attributed to differences in political system, state structure and discursive opportunities. The complete absence of local resistance in East Germany to ERAM prior to German unification contrasts with diverse activities in the West German part of the Morsleben region in the 1980s. However, economic and political 'tipping points' such as German unification in 1990 and the end of the 'Kohl era' in 1998 coincided with rapid changes in the structure and formation of local and regional environmental networks. Almost immediately after the fall of the Berlin Wall, West German environmental groups deployed their expertise and political action repertoires to draw attention to the ERAM case and mobilise support in East Germany. This created some tensions between more established West German environmental groups and activists from the former GDR. This also reflected the complex and oftentimes problematic convergence of East and West German Green movements which continues to date (Jones 1993, Gerber 1999, Rucht 2003). Local initiatives and groups such as Greenkids e.V. formed, some of which developed strong regional and even international links while others focused more on local issues. This suggests that the presence and direction of environmental contestation cannot be solely attributed to (un)favourable political structures. Contextual conditions including the dynamics of local social networks, the presence or otherwise of key activists and geographical proximity play a significant role too.

Mobilities, materialities and the transformations of environmental protest

How useful is the 'new mobilities paradigm' for understanding the trajectories and outcomes of our two cases? An initial focus on more visible mobilities such as the movement of wastes and the circulation of information among activists serves to highlight the complexities of material flows (see Table 3). A cursory examination of our two cases demonstrates the relevance of the types of mobilities outlined in section one. For example, the NGOs and community groups in both cases utilised information and communications technology and traditional media such as newspapers, flyers and radio to coordinate local and regional protests. German and Malaysian NGOs were similarly integrated in global networks, notably Friends of the Earth International, which facilitated exchange of information, sharing of expertise and coordination of campaigns.

But the physical movement of people and objects does not itself determine whether or not spaces become (de-)politicised. Instead, the comparative analysis of mobilisations in Morsleben and Bukit Nenas reveals complex interrelationships between material conditions, political structures, cultural factors and power relations. We must, therefore, move beyond the 'new mobilities paradigm' and emphasise the political nature of mobility. In Germany, unification resulted in changed state structures that allowed for more open contestation and mobilisation. High-profile actions like the spectacular occupation of ERAM (*Morsleben Aktion Stillegung*) by Greenpeace activists in 1993 were complemented by local initiatives such as the mobile exhibition by Greenkids e.V. In contrast, in Malaysia, CAP/SAM's attempts to unite local communities through grassroots mobilisation were hampered by state restrictions. This partly resembled the situation in pre-unification

Table 3.   (Im)mobilities and environmental contestation in Germany and Malaysia.

	Morsleben	Bukit Nenas
Physical movement of objects	• Transport of radioactive waste • Release of pollutants (radiation) • CASTOR	• Separation and transfer of hazardous waste • Release of pollutants (dioxins) • Incineration plant • Movement of investment
Corporeal travel	• (Restrictions on) movement across inner-German border • Travel to/from demonstrations and protests • (In)voluntary immobilisation of anti-CASTOR protesters • Meetings between activists in the region	• (Restrictions on) assembly and protest • Travel of CAP workers to area to mobilise and inform local people • Detention of activists • Policing of protesters • Meetings of international activists
Imaginative and virtual travel	• Internet forums and user groups (e.g. Indymedia community) • Networking of different campaigns	• International networking • Internet forums (e.g. POPs forums) • Networking of different campaigns
Communicative travel	• Use of ICT to inform and mobilise activists and supporters • Media coverage • Mobile exhibition organised by Greenkids e.V.	• Production and dissemination of information (e.g. newsletters) • Local education campaigns • Kualiti Alam website

Morsleben. However, Malaysian campaigners faced additional barriers to mobilisation in the form of ethnic, cultural and religious divisions which had a 'chilling' effect on environmental campaigning.

De-activation of community resistance, including attempts at co-optation, were also features of both cases. In Malaysia, the agenda of development and the promotion of hazardous waste incineration as a 'green technology' and profitable investment enabled the co-optation of locals. In Germany, a recent decision to keep radioactive waste near existing nuclear power plants and avoid the controversial and expensive CASTOR transports exploits the greater acceptance of radioactive risks among people who live near such plants.

What is notable in both cases is the continued importance of the state, but also its contradictory role as 'gatekeeper' of mobilities (of people, waste and information). In the Morsleben case, the East German state sought to simultaneously immobilise public opinion and contain radioactive waste generated by its nuclear energy programme. However, the suppression of dissent proved unsustainable in the long term, contributing to the eventual collapse of the GDR. After unification, transfers from West to East Germany exacerbated the insoluble waste problem, creating a lasting legacy but also providing a continued opportunity for collective action. In Bukit Nenas, attempts by the Malaysian state to regulate hazardous waste coincided with privatisation and the transfer of technology and regulatory frameworks from North to South. However, economic liberalisation was not accompanied by greater political openness. Instead the state effectively demobilised civil society, even as it began to mobilise inward flows of capital and technology to facilitate high-tech treatment for hazardous waste.

The two cases bring us back to a discussion of the inherently political nature of 'wastes on the move'. Contestation occurs whenever waste finally comes to rest: the storage and treatment of waste in specific locales creates opportunities for local and global protest to focus and converge, change direction, or perhaps runs its course and disintegrate (unlike most of the waste). The irreducible nature of radioactive waste in Morsleben contrasts with the partially successful reduction of hazardous waste through high-tech incineration in Malaysia. What matters is not just how mobile they are but also their material characteristics, which are inextricably linked to their political and socio-cultural valency. More toxic and persistent wastes like radioactive waste may have a higher potential for mobilising contestation than wastes that can be reduced through the application of technology. The ERAM case illustrates that nuclear waste retains its mobilisation potential precisely because of its immutably toxic properties. While high-tech incineration in Malaysia produces some toxic residues, they also reduce the bigger problem. The mobilisation potential of incinerator is lessened as the demand for hazardous waste solutions increases.

The two cases illustrate that the effects of political change differ according to the context. In the Malaysian case we can observe a gradual transformation from 'brown authoritarianism' to 'green modernisation' under the auspices of a pragmatic developmental state. The mobilisation of popular contestation is replaced by the mobilisation of technology and of individuals as investors and property owners. Commitment to a high growth model since the late 1980s has led to ever-increasing amounts of waste, pushing the state towards high-tech incineration. This green modernisation process coincided with a discernible shift in public attitudes accepting the introduction of environmental technologies. Shared understandings of environmental threats were weakened once the threats of detention and underlying ethnic tensions were invoked (see Verma 2004, p. 136). The fragmentation of local environmental protest shows the limitations of CAP's bottom-up mobilisation strategy. Despite their continuing opposition to toxics and incineration, they have been left with a technical monitoring role, which is arguably complementary rather than oppositional to high-tech green modernisation. This also shows the relevance of established and newly emerging cultural norms regarding waste and waste management.

The German case, on the other hand, illustrates the importance of distinct political moments or 'tipping points' which radically change the parameters of contestation. The collapse of the Eastern bloc in 1989 and German unification in 1990 led to greater interaction between East and West German green movements, with attendant contradictions. This led to an increase in protest activity around contested sites such as Morsleben but also to the diversification of environmental struggles. The end of the Kohl era and the formation of a red–green coalition government in 1998 initiated a significant shift in nuclear policy which de-prioritised nuclear energy. The *Atomkonsens* restricted the expansion of the nuclear waste problem, but also resulted in a shift in the focus of anti-nuclear protest.

Conclusions: moving beyond the North–South divide: mobilities, transitions and structural transformations

We have explored both the possibilities and limitations of the 'new mobilities paradigm' for analysing environmental contestation, using two case studies from Malaysia and East Germany. By emphasising the materialities of different kinds of hazardous waste, we suggest that they have differing mobilisation potentials. The substantive approach pursued here combines with a focus on mobilities to place the problem of hazardous waste within broader structural and political transformations. The conventional wisdom is that wastes should be immobilised within the country or region of their origin. However, the same conventional wisdom holds that waste treatment technologies should be mobile and their transfer encouraged. The prevailing conceptualisation of technology transfer characterises the relationships in terms of sender and receiver, the latter being seen as passive and in need of development while the former is active and able to assist. The Basel Convention recognises the potential for environmental injustice to result from the movement of hazardous waste from rich industrialised countries or regions to poor and less industrialised ones. This interpretation is also underpinned by Principle Seven of the Rio Declaration concerning 'common, but differentiated responsibility', which sees developed countries as more responsible for environmental degradation and tasks them to provide technology and finance to address global environmental degradation (UNEP 1992). Our substantive focus on intra-national economic and political transitions offers an important corrective to the assumption that hazardous wastes and the technologies to deal with them flow in only one direction – from the industrialised North to the less-developed South.

We noticed that there are at least four problems which become more visible through the lens of mobilities: the assumption that a material disconnection between waste and waste management technology is possible; the belief that less developed countries or regions do not have a waste problem and will continue not to do so as long as they do not receive any imports; the generalisation that the industrialised or developing countries are homogeneously developed or underdeveloped, ignoring internal disparities and transfers; economic growth is not regarded as problematic.

The two case studies take us beyond these assumptions by challenging each of them in turn. Both cases show that a material disconnect between the waste problems and their solutions is impossible. The more developed countries or regions (in these cases, West Germany and Denmark) tend to be the sources of hazardous waste and waste management technologies, and more broadly a mentality of waste management as a governance strategy. However, the inability of the 'more developed' West to deal with its own hazardous waste shows the material limitations of displacement as solution. As regards the belief that less development means no waste problems, both cases illustrate that this is misleading. In Morsleben, a large proportion of the radioactive waste came from West Germany, but East German power plants also produced waste that required disposal. In Malaysia rapid development led to imports of technology and hazardous waste from the North but, although the country is still a net importer of hazardous waste, the proportion of imports is now dwarfed by domestic waste generation.

In relation to the transfer of wastes and associated risks within a country or region, both cases highlight structural inequalities determining the location of waste infrastructure. The Morsleben case shows the structural inequalities inherent *within* the North, coinciding with West–East transfers of environmental bads. In the Malaysian case, the country has begun to converge with the advanced North in terms of waste treatment technology but has experienced high levels of economic development and structural transformation within a fairly authoritarian model. Opportunities for resistance and contention over where wastes are located are circumscribed and the scales are tipped towards state and business interests and against local and disempowered communities.

Finally, the cases cause us to critically question unsustainable models for economic growth. The Malaysian case has followed the green modernisation template, as the government has pursued economic and structural mobility with the ultimate aim of reaching the status of a 'fully developed country' by the year 2020. In pursuit of this vision, it has acquired the latest technologies. However, the expansion of hazardous waste is unsustainable. Despite improved technology, concerns about the toxicity of POPs and the difficulty of monitoring them persist. In the German case, recent statements indicating a possible reversal of the *Atomkonsens* due to increased demands for cheap energy highlight once again the impossibility of dealing with radioactive waste.

According to Sacquet (2005, pp. 48–49), the future of waste is its disappearance. Yet the materiality of waste points to the impossibility of this. Current models of economic development continue to generate unsustainable quantities of waste. We have shown that different wastes do have different mobilisation potentials. While some hazardous wastes can be reduced through the application of 'green' technologies, these treatment methods support unsustainable growth and still produce toxic residues. The longevity and risks associated with nuclear waste mean that they retain their contentiousness. The refusal of waste problems to go away ensures that contestations around waste and waste infrastructure will continue.

Notes

1. Muslim Malays consider pigs to be not only physically but also culturally polluting.
2. CASTOR is an acronym for 'cask for storage and transport of radioactive material'.

References

Basel Convention, 2003. *Malaysia country report*. Basel: Basel Convention Bureau.
Basel Convention, 2006. *Malaysia country report*. Basel: Basel Convention Bureau.
Bastian, U., 1996. *Greenpeace in der DDR*. Berlin: edition ost.
Bauman, Z., 2004. *Wasted lives: modernity and its outcasts*. Cambridge: Polity Press.
Bernama, 2007. Kualiti Alam uses Danish technology to develop its own incinerator. *Bernama Business News,* 25 January.
Beyer, F., 2004. Die (DDR-)Geschichte des Atommüll-Endlagers Morsleben. *Sachbeiträge*, 36. Magdeburg, Germany: Landesbeauftragte für die Unterlagen des Staatssicherheitsdienstes der ehemaligen DDR in Sachsen-Anhalt.
Beyer, F., 2005. Security Policies am Beispiel des Endlagers Morsleben. Unpublished dissertation. Otto-von-Guericke-Universität Magdeburg. Available from http://www.greenkids.de/morsleben/docs/morsleben-sp.pdf [Accessed 22 May 2008].
Blowers, A. and Lowry, D., 1997. Nuclear conflict in Germany: the wider context. *Environmental Politics*, 6 (3), 148–155.
Carson, R., 1962. *Silent Spring*. New York: Houghton Mifflin.
Consumers Association of Penang, 2005. *International POPs elimination project: Malaysia country situation report*. Consumers Association of Penang.
Cresswell, T., 2006. *On the move*. New York: Routledge.
Crouch, H., 1996. *Government and society in Malaysia*. St Leonards: Allen and Unwin.
Cruez, A.F., 2006. A need to tighten loopholes. *New Straits Times,* 22 January.
Fazal, A., 1982. What the consumer movement is about. *Aliran Quarterly*, 3 (3), 66–67.
Federal Ministry for the Environment, Nature Conservation and Nuclear Safety (BMU), 2006. *The closure of the Morsleben repository (ERAM)*. Salzgitter: Federal Office for Radiation Protection. Available from http://www.bmu.de/files/pdfs/allgemein/application/pdf/jc_morsleben.pdf [Accessed 22 May 2008].
Fischer, C. and Boehnke, K., 2004. 'Obstruction galore': a case study of non-violent resistance against nuclear waste disposal in Germany. *Environmental Politics*, 13 (2), 393–413.
Foweraker, J., 1997. Social movement theory and the political context of collective action. *In:* R. Edmondson, ed. *The political context of collective action*. London: Routledge, 64–77.
Gerber, S., 1999. Die Umweltbewegung in der DDR. Unpublished minor thesis. TU Dresden.
Gomez, E.T., 1994. *Political business: corporate involvement of Malaysian political parties*. Kuala Lumpur: Forum.
Gomez, E.T. and Jomo, K.S., 1999. *Malaysia's political economy: politics, patronage and profits*. London: Routledge.
Greenaction/Gaia, 2008. *Burning issues: incinerators in disguise*. Available from http://www.greenaction.org/incinerators/documents/GAIAPresentationIncineratorsInDisguise100907.pdf [Accessed 20 June 2008].

Greenkids e. V., 2008. *Chronik 1995–2004.* Available from http://www.greenkids.de/index.php3?ordner=gk/&seite=chronik.php3 [Accessed 9 July 2008].

Greenpeace, 1997. Strahlende Tropfsteinhöhle: Eine geheime Studie warnt vor Wassereinbrüchen im Atommüll-Endlager Morsleben. *Greenpeace Magazin 3/97.* Available from: http://www.greenpeace-magazin.de/index.php?id=4714&no_cache=1&sword_list[]=morsleben [Accessed 12 June 2008].

Greenpeace, 2003. *Der Fall Morsleben: Ein Atommüll-Endlager bröckelt!* Hamburg: Greenpeace e.V. Available from http://www.greenpeace.de/fileadmin/gpd/user_upload/themen/atomkraft/morsleben.pdf [Accessed 12 June 2008].

Hänel, M., 1998. 'Das Ende vom Ende': Zur Rolle der Energiewirtschaft beim Systemwechsel 1980–90. *Occasional Papers in German Studies,* No. 15. Edmonton, Canada: University of Alberta. Available from http://www.nuklearezukunft.de/ddratom/ddratom.pdf [Accessed 12 June 2008].

International Bar Association, *et al.*, 2000. Justice in jeopardy: Malaysia 2000. *Report on behalf of the International Bar Association, the ICJ Centre for the Independence of Judges and Lawyers and the Commonwealth Lawyers' Association and the Union Internationale des Avocats.* Available from http://www.ibanet.org/images/downloads/malaysia.pdf [Accessed 12 July 2008].

Jones, M.E., 1993. Origins of the East German environmental movement. *German Studies Review,* 16 (2), 235–264.

Keck, M. and Sikkink, K., 1998. *Activists beyond borders.* Ithaca, NY: Cornell University Press.

Lindemann, I., 2007. Atommüllendlager Morsleben – ein einsturzgefährdetes Endlager vor der Stillegung. *Strahlentelex* No. 482/3, 6.

Loh, F. and Khoo, B.T., 2002. *Democracy in Malaysia: discourses and practices.* Richmond: Curzon Press.

Meadows, D.H., 1991. *The global citizen.* Washington, DC: Island Press.

Means, G.P., 1991. *Malaysian politics: the second generation.* Kuala Lumpur. Oxford University Press.

Rootes, C., 1997. Shaping collective action: structure, contingency and knowledge. *In*: R. Edmondson, ed. *The political context of collective action.* London: Routledge, 81–104.

Rootes, C., 2007. Acting locally: The character, contexts and significance of local environmental mobilisations. *Environmental Politics,* 16 (5), 722–741.

Rucht, D., 1990. Campaigns, skirmishes and battles: Anti-nuclear movements in the USA, France and West Germany. *Industrial Crisis Quarterly,* 4, 193–222.

Rucht, D., 2003. Neue Soziale Bewegungen. *In*: U. Andersen and W. Woyke, eds. *Handwörterbuch des politischen Systems der Bundesrepublik Deutschland.* 5th ed. Opladen: Leske + Budrich, Lizenzausgabe Bonn: Bundeszentrale für politische Bildung (on-line). Available from http://www.bpb.de/wissen/0618292007505209 2624361933117076,0,0,Neue_soziale_Bewegungen.html [Accessed 23 June 2008].

Sacquet, A., 2005. *World atlas of sustainable development.* London: Anthem Press.

Salleh, I. and Meyanathan, S., 1993. Malaysia: growth, equity and structural transformation. *Occasional Paper Series.* Washington, DC: World Bank.

Scally, D., 2008. Waste storage issue continues to dog German nuclear debate. *Irish Times,* 26 July, 9.

Standing, G., 1993. Labour flexibility and the Malaysian manufacturing sector. *In*: K.S. Jomo, ed. *Industrialising Malaysia: policy, performance, prospects.* London: Routledge, 40–76.

Third World Network, 1988. *Toxic terror: dumping of hazardous wastes in the Third World.* Penang: Third World Network.

Third World Network, 2001. *International environmental governance: some issues from a developing country perspective.* Penang: Third World Network.

UNEP (United Nations Environment Programme), 1992. Rio declaration on environment and development. *UN Conference on Environment and Development*, 3–14 June, Rio de Janeiro. Available from http://www.unep.org/Documents. Multilingual/Default.asp?DocumentID=78&ArticleID=1163 [Accessed 7 July 2008].

UNEP (United Nations Environment Programme), 2004. *State of waste management in Southeast Asia*. Available from http://www.unep.or.jp/Ietc/Publications/spc/ State_of_waste_Management/7.asp [Accessed 18 June 2008].

Urry, J., 2000. *Sociology beyond societies: mobilities for the twenty-first century*. London: Routledge.

Urry, J., 2007. *Mobilities*. Cambridge: Polity.

Verma, V., 2004. *Malaysia: state and civil society in transition*. Petaling Jaya, Malaysia: Lynne Rienner/Strategic Information Research Development.

Index

Gordon, C.: and Jasper, J. 39
Grand Bois oilfield: Louisiana 36, 43-4
Grassroots Recycling Coalition 30
Greece 11, 12, 16; Attica landfill sites 129-
38; ECJ compliance violations 129, 130;
grassroots mobilisations 123-43; landfill
sites 128-30; municipal waste disposal
127-9; NIMBY arguments 124-5, 126,
127, 130-8; packaging waste 128; Waste
Framework Directive 129
Greenpeace 27, 57, 58, 67, 73, 154, 156,
157
grievance framing: outcome determinants
38-9

Hannoversche Allgemeine Zeitung 155
hazardous waste: movements and mobilities
144-64; politics 144-64
Health Protection Agency (HPA) 62
health risks: incineration 6-10
Hellenic Recovery Recycling Corporation
(HERRCO) 128
Hogg, D. 70
Housing Authority of New Orleans
(HANO) 44

incineration: ash 28; British campaigns
53-79; changing technologies 14-16;
combined heat and power (CHP) 14,
70; dioxins 7; efficiency 14; emission
toxicity 6, 7; epidemiological studies
6; France 101-22; French biodiversity
issues 111-14; Friends of the Earth 57-8,
61; Greenpeace 57, 58; health risks 6-10;
and incinerator siting 9-10; Ireland 80-
100; particulates 7-8; public fears 7; UK
campaign success determinants 59-66;
United States campaigns 25-9; waste
5-10
Institute for Local Self Reliance (ILSR) 28
International POPs Elimination Project
(IPEN) 152
International Solid Waste Association
(ISWA) 105
Ireland 11, 16; An Board Pleánala (planning
board) 85, 95; anti-incineration
campaigns 80-100; Cork Harbour
Alliance for a Safe Environment
(CHASE) 93-6; DuPont plant 84, 86-
92; Galway for a Safe Environment
(GSE) 92-3; and governing waste
81-2, 83; Green parties 97-8; National

Development Plan 83; North-South
governance context 85-6; peace and
growth 96-9; state governance context
82-5; and Westminster governance
model 82

Jackson, L. 49
Jasper, J.: and Gordon, C. 39
Jefferson, W. 45
Jordan, J.: and Gilbert, N. 91

Karamichas, J.: and Botetzagias, I. 12
Kasperson, R.E. 123
Kemberling, M.: and Roberts, J.T. 12
Khoo, S-M.: and Rau, H. 13
Kitschelt, H. 39
Kousis, N. 132
Krohn, W.: and Van den Daele, W. 5
Kualiti Alam 148, 150, 151-2

Laliotis, K. 130
Landfill Allowance Trading Scheme
(LATS) 68
landfill sites 4-5, 12, 24, 26-7, 29; EU
Landfill Directive 55; France 102-3;
Greece 128-9; United Kingdom 54-5,
67-8; USA 125
Landfill Tax: UK 67-8
legal strategy: and outcome determinants
37-8
Leonard, L. 108
Louisiana: Agriculture St landfill site 36,
44-6; Citizens Against Nuclear Trash
(CANT) 40-1; environmental justice
outcome determinants 35-52; Grand
Bois oilfield contamination case 36,
43-4; PVC factory siting dispute 36, 37,
41-2; St. James Citizens for Jobs and
the Environment (SJCJE) 42; uranium
enrichment plant siting dispute 35-6,
37, 40-1
Louisiana Energy Services (LES): plant
siting dispute 35-6, 37, 40-1, 46, 47, 50
Louisiana Environmental Action Network
(LEAN) 47
Love Canal: environmental justice
movement 24-5, 25, 27

McAdam, D. 37, 38
McCann, M. 37
Malaysia: anti-toxics campaigns 145,
146, 147, 149-53, 157-8; Bukit Nenas

Citizenship Studies

CHIEF EDITORS:

Professor Engin F. Isin, *POLIS, The Open University, UK*
Professor Bryan S. Turner, *Wellesley College, USA*

Citizenship Studies publishes internationally recognised scholarly work on contemporary issues in citizenship, human rights and democratic processes from an interdisciplinary perspective covering the fields of politics, sociology, history and cultural studies. It seeks to lead an international debate on the academic analysis of citizenship, and also aims to cross the division between internal and academic and external public debate.

The journal focuses on debates that move beyond conventional notions of citizenship, and treats citizenship as a strategic concept that is central in the analysis of identity, participation, empowerment, human rights and the public interest. Citizenship is analysed in the context of contemporary processes involving globalisation, theories of international relations, changes to the state and political communities, multiculturalism, gender, indigenous peoples and national reconciliation, equity, social and public policy, welfare, and the reorganisation of public management.

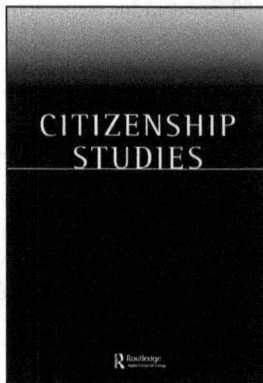

CITIZENSHIP STUDIES

Environmentalism in the United States
Changing Patterns of Activism and Advocacy

Edited by **Elizabeth Bomberg**, University of Edinburgh, UK
and **David Schlosberg**, Northern Arizona University, USA

Routledge Paperbacks Direct

Environmentalism – defined here as activism aimed at protecting the environment or improving its condition – is undergoing significant change in the United States. Under attack from the current administration and direct questioning from its own ranks, environmentalism in the US is at a crossroads. This title explores the changing patterns of and challenges to environmentalism in the contemporary US. More specifically, it will examine the following dynamics:

· the re-conceptualisation of core ideas and strategies defining US environmentalism;
· questions of identity and relations with other advocacy groups (including labour, global justice and women's groups)
· institutional change (especially the shift away from regulatory policies and approaches)
· the expanding arenas of activism, to both above and below the state
· environmentalists' response to Bush administration policies and priorities.

Contents

Paperback
ISBN
978-0-415-48394-0
£20.00

Order this book today at:
www.routledge.com/paperbacksdirect

Routledge
Taylor & Francis Group

Routledge
Taylor & Francis Group

Local Government Studies

EDITOR:

Colin Copus, *De Montfort University, UK*
Philip Whiteman, *University of Birmingham, UK*

2007 Impact Factor: 0.486
56/93 (Political Science), 20/27 (Public Administration)
© 2008 Thomson Reuters, *Journal Citation Reports®*

Local Government Studies is the leading journal for the study of the politics, administration and management of local affairs. The journal publishes articles which contribute to the better understanding and practice of local government and which are of interest to scholars, policy analysts, policymakers and practitioners. The focus of the journal is on the critical analysis of developments in local governance throughout the world. The editors particularly welcome studies of issues related to European local government. *Local Government Studies* provides a unique forum for the consideration of all issues related to sub-national levels of government.

To sign up for tables of contents, new publications and citation alerting services visit www.informaworld.com/alerting

e-updates
Taylor & Francis Group

Register your email address at www.tandf.co.uk/journals/eupdates.asp to receive information on books, journals and other news within your areas of interest.

Powered by
informaworld

For further information, please contact Customer Services at either of the following:
T&F Informa UK Ltd, Sheepen Place, Colchester, Essex, CO3 3LP, UK
Tel: +44 (0) 20 7017 5544 Fax: 44 (0) 20 7017 5198
Email: subscriptions@tandf.co.uk

Taylor & Francis Inc, 325 Chestnut Street, Philadelphia, PA 19106, USA
Tel: +1 800 354 1420 (toll-free calls from within the US)
or +1 215 625 8900 (calls from overseas) Fax: +1 215 625 2940
Email: customerservice@taylorandfrancis.com

View an online sample issue at:
www.tandf.co.uk/journals/lgs

For Product Safety Concerns and Information please contact our EU
representative GPSR@taylorandfrancis.com
Taylor & Francis Verlag GmbH, Kaufingerstraße 24, 80331 München, Germany